# linotronic®
## imaging
### handbook

the desktop
publisher's guide
to high-quality
text and images

james cavuoto + stephen beale

**MICRO PUBLISHING PRESS**
Torrance, California

Linotronic Imaging Handbook
The Desktop Publisher's Guide to High-Quality Text and Images
James Cavuoto and Stephen Beale

**Published by:**
Micro Publishing Press
21150 Hawthorne Blvd., Suite 104
Torrance, CA 90503
(213) 371-5787

Copyright © 1990 by Micro Publishing

First Printing, November, 1990

Printed in the United States of America

Library of Congress Catalog Card Number: 90-92100
ISBN 0-941845-06-0

This book is designed to provide information about Linotronic imagesetters. Every effort has been made to make this book as complete and accurate as possible. But no warranty of suitability, purpose, or fitness is implied. The information is provided on an "as-is" basis. The publisher shall have neither liability nor responsibility to any person or entity with respect to any loss or damages in connection with or arising from the information contained in this book. This book is an independent publication of Micro Publishing Press. It was developed with the permission of Linotype AG, but Linotype AG assumes no responsibility for the content of the book.

Many of the designations used by manufacturers and sellers to distinguish their products are claimed as trademarks. Where these designations appear in the book and the authors were aware of a trademark claim, the designations have been printed with initial capital letters—for example PostScript.

Linotype and Linotronic are registered trademarks of Linotype AG, and/or its subsidiaries.
PostScript and Adobe are registered trademarks of Adobe Systems, Inc.
Separator, Streamline, Photoshop, and Illustrator are trademarks of Adobe Systems, Inc.
Macintosh, LaserWriter, and AppleTalk are registered trademarks of Apple Computer Inc.
IBM and IBM PC are registered trademarks of IBM Corp.
Aldus, PageMaker, and Aldus FreeHand are registered trademarks of Aldus Corp.
ColorStudio, ImageStudio, and DesignStudio are trademarks of Esselte Pendaflex Corp.
Pantone and PMS are registered trademarks of Pantone Inc.
All other company and product names are trademarks or registered trademarks of their respective owners.

# CONTENTS

# PREFACE

As editors of *Micro Publishing News*, the Southern California desktop publishing newspaper, we became aware of many of the issues confronting users and service bureaus. We also learned firsthand the challenges that typical desktop publishing users encounter during a publishing project such as this. This entire book, including the cover and color insert, was produced with desktop publishing products and output on Linotronic imagesetters.

A variety of tools made up our repertoire. We created text files with Microsoft Word, both the Macintosh and IBM versions. Graphic images, including the cover, were produced in Aldus Freehand, Adobe Illustrator, SuperPaint, DeskPaint, and DeskDraw on the Macintosh. We used Photoshop, ColorStudio, ImageStudio, and Digital Darkroom to touch up images scanned on the Hewlett-Packard ScanJet Plus, Microtek 300ZS, XRS 3c, or Barneyscan Slide Scanner. Radius' Screen Capture utility allowed us to capture 24-bit color images. We used TOPS to transfer graphic images created or captured on the IBM PC. PrintBar software from Bear Rock Technologies produced the bar code on the back cover.

We used Aldus PageMaker version 4 running on a Macintosh IIcx for page layout. The black-and-white pages were output on a Linotronic 300 imagesetter with RIP II; halftone screen was 100 lpi. Color pages were separated with Aldus PrePrint and output on a Linotronic 330 with RIP 30 at RPI, a service bureau in Los Angeles. The color photographs were produced with a 133-line screen.

This book was printed on a web press at Delta Lithograph in Valencia, CA. The paper stock is a 60-lb. offset. The four-page color insert is printed on 70-lb. coated stock.

# ACKNOWLEDGMENTS

We would like to thank several individuals and organizations who helped us produce this book. First and foremost, Nancy Carr at Linotype deserves credit for her dedication to this project and her insightful advice. Jim Hamilton, Craig Norris, Jimmy Yellen, Rich Vitale, Ricky Schrieber, and Gus Keysor at Linotype also offered valuable advice. Manfred Werfel at Linotype in Germany was very helpful, as were George Barber, Michael Barry, and Terry Kowalski in the Los Angeles office.

Howard Fenton contributed a major portion of Chapter Eight, and Ron McPherson of From Art to Design in Los Angeles provided overall art direction as well as the cover design. Several representatives of Los Angeles-area service bureaus offered assistance, including Rick Valasek of Quicktype & Design, Randy Small of Vision Graphics, Galen Anderson of Anderson's Typesetting, Bill Swann of RPI, and Sharon Davis of Wheeler/Hawkins. Derek Handova and Mary O'Neill at Micro Publishing Press offices performed research and proofreading tasks.

We wish to thank the following companies that loaned us hardware we used to produce this book: Linotype–Lintronic 300 imagesetter; Radius – Color Display and DirectColor/24 interface; QMS–ColorScript 100 Model 10 printer; XRS – 3c scanner; Apple Computer – Personal LaserWriter NT; Umax–UG80 scanner; and Barneyscan – Slide Scanner.

The following software companies were cooperative in sending us programs and technical information we needed: Adobe Systems, Aldus Corp., Astral Development, Letraset, Symsoft, Ultimate Technographics, and Ventura Software.

# CHAPTER 1

# Introduction to Linotronic Imaging

The Linotronic imagesetter is hardly a household word, and certainly not the kind of product you'd see advertised on television. But in its own way, this complex device has touched the lives of millions of people around the globe who depend on the written word. If you've done much reading during the past couple of years—books, magazines, brochures, or anything else—chances are that you've seen the work of a Linotronic imagesetter.

The Linotronic imagesetter is a sophisticated output device manufactured by Linotype AG of Eschborn, West Germany. It uses state-of-the-art laser and computer technology to produce high-quality type and graphics on film or resin-coated paper. But its impact goes beyond its technical features, because it also represents a bridge between old and new in the publishing business.

In the old days, publications were produced by means of bulky, expensive typesetting systems run by specially trained operators. Creating a page was a slow and costly affair. Even when typesetting was finished, a production artist had to paste the type onto a layout board along with illustrations and other page elements. The whole process could take days under the best circumstances. If any mistakes were found, it was back to the typesetting machine for corrections.

# The Origin of Desktop Publishing

In the mid-1980s, a computer application known as "desktop publishing" changed all this. For less than $10,000, you could purchase a personal computer, laser printer, and page composition software and become your own personal publishing company. Instead of pasting galleys onto a layout board, you could compose a page, complete with text and graphics, on a computer screen. Instead of sending your work to a typesetting house, you could quickly and easily produce it yourself on a laser printer. Just as Gutenberg's invention of the printing press brought books and other publications to the masses, desktop publishing made it possible for ordinary people to produce their own professional-looking documents.

There was just one problem in this rosy picture. The inexpensive laser printers that make desktop publishing possible offer a limited print density—otherwise known as resolution—of 300 or 400 dots per inch (dpi). This is satisfactory for many kinds of documents, but other publications have strict quality requirements that can only be met by "typeset-quality" resolution, generally regarded as 1200 dpi or more. For these projects, 300-dpi resolution just doesn't cut it.

Enter the Linotronic imagesetter (Figure 1-1). The imagesetter uses a combination of computer and phototypesetting technology to produce documents at resolutions ranging to more than 3000 dpi. Because of a powerful piece of printing software known as PostScript®, the imagesetter can produce documents created by inexpensive desktop publishing systems. And, true to its name, the Linotronic imagesetter can produce images—even photographic halftones—in addition to text. You can compose a page with text and graphics on your computer and have it printed on the imagesetter without any need for paste up.

**Figure 1-1.** The Linotronic 300 (left) and 500 (right) imagesetters.

PostScript, one of the key components of the Linotronic imagesetter, is a page description language developed by Adobe Systems Inc. of Mountain View, California. A page description language is a set of commands that tell a laser printer or imagesetter exactly how to create a page. PostScript is the most popular page description language for microcomputer applications because of its association with Apple's Macintosh computer. But users of other computers can also take advantage of PostScript.

A major advantage of PostScript is a feature known as "device independence." If you create a document for output on one PostScript printer, it can be produced with little or no modification on any other PostScript output device, including a Linotronic imagesetter. Along with this goes a property known as "resolution independence." When you print your document on a PostScript laser printer, its resolution might be 300 or 400 dpi. But when you print the document on a Linotronic imagesetter, its resolution can be anything from 1270 to 3386 dpi, depending on the particular model you are using and the resolution settings you have chosen. The page, rather than being "locked into" a particular resolution, can be printed at the maximum density of the output device.

PostScript, as we shall see later, offers many benefits beyond device independence. But this particular feature is responsible for much of the imagesetter's success. Like any piece of typesetting equipment, the Linotronic imagesetter is not inexpensive. Even the least expensive model is well beyond the price range of most desktop publishers. But many desktop publishing service bureaus offer PostScript output on their Linotronic imagesetters. You can design a publication on an inexpensive microcomputer and have it produced at high resolution for as little as $5 to $10 per page. Before you go to the service bureau, you can print the publication on your own PostScript laser printer to check for any mistakes. You thus get all the advantages of desktop publishing along with the benefits of high-resolution output.

It sounds simple, and in many cases it is. But working with a sophisticated piece of equipment like a Linotronic imagesetter can often be a challenge. You might find that the typefaces you specified when creating your publication don't appear as you expected. Lines that looked OK on a laser printer might be too thin at higher resolutions. Halftone photographs might look too dark or too light. What's most disturbing, if you're a service bureau customer, is that you're probably paying for this faulty output.

Most publishers who use the Linotronic imagesetter have high quality requirements. They are paying a premium to get everything just right: the type, the lines, the pictures, everything. For these users, it makes a lot of sense to learn about techniques for getting the best possible output from their imagesetters.

That's what this book is all about. As authors and consultants in the field of desktop publishing, we have learned to appreciate the benefits—and challenges—of the Linotronic imagesetter. We have marveled at the quality of its output, and cursed when a section of text or graphics did not turn out the way we wanted. Above all, we have seen that it takes a healthy dose of knowledge to harness the power of this complex device.

In this book, we'll try to provide this knowledge. We'll describe the technology behind the Linotronic imagesetter, but that is not our primary aim. Instead, we want to provide the tools you'll need to produce high quality text and graphics on the Linotronic device. Beyond describing the Linotronic hardware, we'll discuss the software—publishing software, graphics software, font software, system software —that gives the imagesetter its real power. And most important, we'll discuss how this software can be used most effectively.

## Linotronic Product Line

So far, we have described the Linotronic imagesetter as a single entity. But there are actually five models as of this printing: the Linotronic 200SQ, Linotronic 300, Linotronic 230, Linotronic 330, and Linotronic 530. Older models of the Linotronic imagesetter, such as the Linotronic 100, can still be found in service bureaus and corporate publishing departments. The models vary in their resolution options and other features, but all are capable of producing high-quality text and graphics.

The Linotronic imagesetter is not the only high-resolution, PostScript-based output device. But it was the first, and remains by far the most widely installed PostScript imagesetter. "Linotronic" is a registered trademark of Linotype, and the term should not be used to describe imagesetters from other manufacturers. Linotype Company specifies that the product should be referred to as the "Linotronic imagesetter" instead of the plain old "Linotronic" (or "Lino"). We will observe this usage throughout the book.

# Focus of the Book

Although we will cover all current Linotronic models, our focus is on the PostScript-based devices. Linotronic imagesetters can actually speak several languages other than PostScript, but we are not terribly interested in these. For most desktop publishers, it is PostScript that provides the gateway to imagesetting. We'll provide brief descriptions of other formats like Linotype's CORA, but the bulk of the book focuses on PostScript and its applications.

We'll follow this with descriptions of several "front-end" environments that allow you to create pages for imagesetter output. We'll go on to discuss the complexities of fonts and font management. We'll describe what is meant by an "Encapsulated PostScript" file and explain its use. We'll discuss the issues involved in image scanning, especially scanning of photographic images. We'll provide tips about how to best reproduce documents created on an imagesetter. And we'll offer tips about how to work with a Linotronic service bureau.

But first, let's take a look at the history behind the Linotronic imagesetter. Then we'll peek below the hood of this revolutionary device.

# Linotype and the History of Typesetting

The Linotronic imagesetter is the latest in a series of Linotype products that go back to the first typesetting machine ever invented. A history of Linotype is thus a history of typesetting itself. It begins in the late 1800s with an ingenious German immigrant named Ottmar Mergenthaler (Figure 1-2), who created the biggest innovation in publishing technology since Gutenberg invented the printing press.

**Figure 1-2.** Ottmar Mergenthaler, inventor of the Linotype.

If Ottmar Mergenthaler were alive today, we can easily imagine him designing computer hardware in the Silicon Valley. Born in 1854 in Hachtel, Germany, he showed an early mechanical aptitude and became an apprentice at age 14 in his uncle's watchmaking shop. Four years later, he moved to the United States and joined his cousin in a small Washington, D.C. company that manufactured prototypes of new inventions. There, he met a famed court reporter named J.O. Clephane, who was seeking an efficient means of producing pages.

At the time, printing presses had developed to the point where newspapers and other publications could be reproduced quickly and cheaply at high volumes. But the pages themselves were composed in a slow, laborious process in which workers manually selected and placed small metal casts containing individual characters. Clephane was one of many people in the printing industry seeking a more productive method of setting type.

Clephane hired Mergenthaler's shop to produce an invention that would automate the creation of printing plates. Ottmar thus began work on a device that would lead to the first true typesetting machine. Along the way, Clephane and a wealthy Washington attorney named L.G. Hine organized a group of investors to provide financial support for development of the invention. The venture, known as the National Typographic Company, was the precursor to the modern Linotype Company.

In 1886, a group of powerful newspaper owners headed by New York Tribune publisher Whitelaw Reid became majority investors in the company, and its name was changed to the Mergenthaler Printing Company. Ironically, Mergenthaler himself held minimal stock in the company and had little influence over corporate decisions aside from his role as the chief inventor. His original supporters, Clephane and Hine, remained as directors.

## The First Linotype Machine

On July 3, 1886, after years of false starts and design improvements, Ottmar Mergenthaler gave his new invention its first real-life test at the offices of Reid's New York Tribune. It was a complicated device that looked like a nightmarish version of a pipe organ (Figure 1-3). Driven by steam, it had a typewriter-like keyboard along with a complex series of pipes, gears, and cables. The type itself was stored as metal casts—one for each character—in a removable magazine. As an operator hit keys on the keyboard, the casts slid down a chute and fell into place on a line bar. A slug containing hot metal was pressed against the bars, leaving impressions of the character shapes. Type was thus composed line by line until the entire page was produced.

**Figure 1-3.** The original Linotype machine.

The Blower

It may seem cumbersome by today's standards, but in 1886 the machine was regarded as a major technological achievement. Observers at the demonstration applauded as Mergenthaler sat down to produce the first slug of metal type. As it came out of the machine, Reid held it up and reportedly said, "Ottmar, you've done it. A line of type." From that statement the name "Linotype" was born (or so the story goes).

Mergenthaler's invention received wide publicity, and an earlier version had even been demonstrated for President Chester Arthur. But Mergenthaler, ever the engineer, ran into constant disagreements with the investors, who wanted to produce the machine as a commercial product before he thought it was ready. Twelve Linotypes were originally produced, and Ottmar wanted to give press operators a chance to use them for a while before manufacturing mass quantities. This process would give him a chance to iron out any bugs in the machine. But the investors were eager to recoup their money as soon as possible and ordered him to make 100 machines.

The disagreements continued and Ottmar eventually left the company, selling his stock in the process. Strapped for funds, he was still able to design an improved version of the Linotype called the Simplex. He sought out his original financial supporters, Hine and Clephane, for money that would enable him to manufacture the new machine. From them he discovered that a management shakeup was imminent at the company that still bore his name. In 1889, Whitelaw Reid was deposed as the company president in favor of Hine. A year later it was reorganized as the Mergenthaler Linotype Company, with the Simplex its major product. By 1892, 1000 Linotypes had been sold or rented, and a year later the machine was a popular attraction at the World's Fair in Chicago.

Ottmar Mergenthaler died in 1899, but the company that was his namesake lived on. By then, the Linotype had revolutionized the publishing industry, causing a three-fold increase in the number of newspapers and other publications available to the public. By 1904 about 8000 Linotypes were installed in the United States. That same year, a group of publishers meeting in New York estimated that the machine allowed one operator to do the work of four hand composers.

## The Emergence of Phototypesetting

Linotype's original patents eventually expired, and several competitors arose. But the company prospered due to its well-established presence and its collection of typefaces, known as the Mergenthaler Type Library. Linotype and its rivals continually enhanced the hot-metal line casting machines, but Mergenthaler's basic design was little changed. It remained so until the 1950s, when a new technology called phototypesetting emerged.

The first phototypesetters replaced the metal casts used in the original linecasters with pieces of negative film containing character shapes. Light was shined through the character shapes onto photosensitive paper. Strips of type could then be pasted into galleys and used as originals for offset reproduction. Subsequent generations of phototypesetters used a cathode ray tube (CRT) to etch character shapes on photosensitive paper, but the general idea remained the same.

Linotype was not the first company to develop a phototypesetter, but it quickly adapted its products to the new technology. In the 1960s and '70s the company introduced numerous phototypesetters that offered the advantage of using the same typefaces found in the hot metal machines. Then, in 1979, the company introduced a revolutionary digital typesetter called the Omnitech 2000. This product, which combined typesetting with computer technology, employed laser imaging and digital fonts to generate

**Figure 1-4.** A front-end workstation.

pages. The product was not much of a commercial success, but it was a precursor to the Linotronic imagesetter, which also uses laser imaging technology.

The company also employed computer technology in the development of terminals known as "front-end workstations" (Figure 1-4). These devices, which include a keyboard and video monitor, allow an operator to compose pages that are then sent to the typesetter for output. The Linotronic imagesetter can be used with some of these front-ends, but a Macintosh or IBM-compatible microcomputer can also be used for the same purpose.

In 1983, Linotype signed an agreement with Adobe Systems that would allow Adobe to offer typefaces from the Linotype Type Library with PostScript-based laser printers. As part of the deal, Linotype received the right to offer PostScript compatibility in its Linotronic imagesetters. Users of desktop publishing systems would have their bridge to high-resolution output capabilities.

For several years, the Linotronic imagesetter was the only high-resolution output device to offer PostScript capabilities, and it quickly became a standard item at desktop publishing service bureaus. Several competitors of Linotype now offer their own PostScript imagesetters. Some of these companies use the version of PostScript developed by Adobe Systems, while others use PostScript "clone" languages that emulate the functions in PostScript. But Linotronic imagesetters remain by far the most popular high-resolution PostScript devices.

Now that we have a background on Linotype the company, let's learn a bit more about its machines. In the following chapter, we'll take a look inside a Linotronic imagesetter to see how it operates.

# CHAPTER 2

# Inside the Linotronic Imagesetter

The Linotronic imagesetter is a far cry from the original line-caster invented by Ottmar Mergenthaler. From all outward appearances, it looks like a simple machine, a large metal cabinet with a few push-button controls compared with the original steam-driven Linotype and its maze of mechanical parts. Inside, however, the imagesetter reveals itself to be a complex technological undertaking.

## Elements of a Linotronic Imagesetter

A PostScript imagesetter has three essential elements aside from the computer on which pages are composed: the raster image processor (RIP), the imager, and the film processor. The actual images are produced by the imager, but without the RIP, there is no way to get image data into the device. And without the film processor, there is no way to see the results of the imagesetter's output.

### The Raster Image Processor

The RIP (Figure 2-1) is not the most obvious element in the imagesetter, but it may be the most important. The function of a RIP is to convert page description instructions from a desktop publishing system into a format that can be accepted by the imagesetter. It thus acts as a translator between the microcomputer user and the high-resolution output capability of the imagesetter. This translation re-

**Figure 2-1.** The RIP (left) produces output for the Linotronic imagesetter.

quires intensive processing on the part of the RIP, which tends to slow the output speed. A faster RIP thus adds to the ultimate performance of the imagesetter.

The Adobe PostScript RIP used with new Linotronic imagesetters comes in one of two varieties, the RIP 4 and the RIP 30. There were three RIP versions that preceded these two current products. The RIP 1 was sold with the original Linotronic 100 models. RIP 2, also known as the Atlas RIP, is a faster version of RIP 1 and continues to be found in many Linotronic imagesetters. RIP 3, also known as the Atlas Plus, is faster still than RIP 2. It is particularly useful for processing photographic halftones and other images, which tend to contain a lot of data.

In the Linotronic 300, 330, and 530, the RIP is a separate unit that looks like a desktop computer set on its side. In the Linotronic 200SQ and 230, the RIP is built into the imagesetter. But whether separate or built-in, all RIPs have common elements. All have a set of hardware connections that permit communication with computers and other equipment. The "AppleTalk" port allows connection with Apple's Macintosh computer and other computers using Apple's local-area networking system. The RS-232 serial and Centronics parallel ports are standard interfaces used by IBM-compatible computers. The newer versions of the RIPs also include an "Ethernet" port, which allows their use with computer networks using the popular Ethernet LAN.

The RIP also includes a hard disk and internal memory (also known as RAM), which are used to store typefaces and other data. A typical RIP might contain 40 to 80 megabytes of hard disk space, enough to store 40 to 80 million characters of information. Any data stored on the disk remains whether the machine is turned on or off. Data can also be stored in the 3 to 5 MB of RAM in the RIP, but this is lost when power is turned off.

The "brain" of the RIP is a 68020 microprocessor manufactured by Motorola, the same chip used in Apple's Macintosh II computer. It converts page description commands in the PostScript language into an image that is then sent to the imagesetter. The image is built as a series of tiny dots as small as one-three-thousandth of an inch (depending on the resolution setting and the capability of the machine).

## The Laser Imager

The imagesetter itself is the bulkiest part of the system, weighing 400 lbs. in the case of the Linotronic 300 and occupying about as much space as a medium-sized photocopier. It is connected by data cable to the RIP, from which it receives the image of a page in the form of tiny dots. This image is converted into a series of signals transmitted to a laser, which sends a thin beam of light through a series of lenses and mirrors, etching the dots onto photosensitive paper or film (Figure 2-2).

The film or paper are found in two cartridges on top of the imagesetter. The input cartridge contains unexposed film or paper, while the take-up cartridge contains film that has been exposed to the laser beam. Controls on the front of the imagesetter allow you to advance the film and cut it so the take-up cartridge can be removed. Controls on the top of the

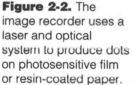

**Figure 2-2.** The image recorder uses a laser and optical system to produce dots on photosensitive film or resin-coated paper.

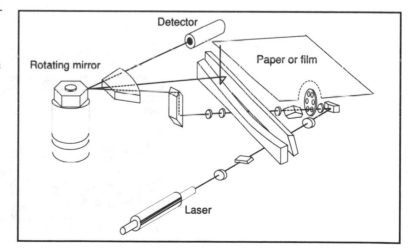

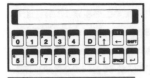

**Figure 2-3.** An LCD panel on the top of the imagesetter allows the operator to choose output resolution and other options.

imagesetter allow you to set the desired resolution—630, 1270, 1693, 2540, or 3386 dpi, depending on the model. An LCD panel on top of the unit displays messages about the machine's operating status and condition (Figure 2-3).

## The Film Processor

Once a print job has been completed, the film is cut and the take-up cartridge removed. The film or paper is then run through a processor (Figure 2-4), which applies a series of chemicals similar to those used by photo processing labs. The first chemical the film encounters is the developer. As with traditional photography, the developer darkens only the areas of the film that have been exposed to the laser. Next, the film passes through a bath containing fixer, which chemically affixes the blackened area of the output film. Finally, the film or paper passes through a water bath, which washes away any remaining chemicals. Before exiting the processor, the paper or film passes through a heating element, which dries any remaining moisture.

The final product is a roll of paper or film containing one or more pages. Since paper or film for a Linotronic imagesetter comes on a continuous roll, the operator must cut each

**Figure 2-4.** The film processor "develops" the film produced by the imagesetter in much the same way a photo processing lab develops snapshots.

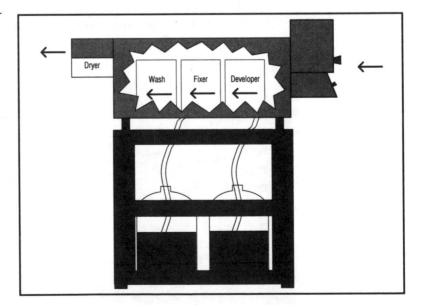

individual page with a paper cutter or scissors. Individual pages can also be trimmed to the desired size, such as 8 1/2 by 11 inches.

# Light Sources

The five Linotronic models differ in the way some of these features are implemented. The Linotronic 300, 330, and 530, for example, use a helium-neon laser to etch dots on film or paper. Though expensive, this type of laser can produce a very small dot size—just 20 microns. Another advantage is that it produces red light in the visible spectrum, which allows its use with high-quality film media. It can also place dots on pages up to 18 inches wide. The Linotronic 200SQ and 230, on the other hand, use a less expensive diode laser, similar to those used in many laser printers. Its dot size is slightly larger than what a helium-neon laser creates, reducing the maximum print density it can offer. On the plus side, it requires less power than a helium-neon laser and is thus smaller and less expensive. Its low cost allows Linotype to keep the 200SQ in an affordable price range.

# Resolution

We have already noted that the Linotronic 200 and 230 have built-in RIPs as opposed to the stand-alone RIP in other Linotronic models. They are also more compact, with simpler paper paths, than the 300 or 330. But for many users, the key distinction among the different models is resolution. The Linotronic 100, now discontinued, offered maximum resolution of 1270 dpi. The Linotronic 200 offers 633-, 1270-, and 1693-dpi resolution. The Linotronic 300, 500, and 530 offer 633-, 1270-, and 2540-dpi resolution. The Linotronic 330 offers up to 3386-dpi resolution. The 500 and 530 differ from the 300 and 330 in that they offer a larger page size suitable for use in newspaper production.

Pages produced on the 300 and 330 can measure up to 11.7 inches wide, but the 500 and 530 produce pages up to 18 inches wide.

As we shall see in later chapters, it is not always necessary or desirable to produce pages with the highest possible resolution. Many documents look fine at 1270 or 1693 dpi, and the lower resolutions can save you quite a bit of expense if you use a service bureau for Linotronic output.

But before we cover resolution settings, we need to discuss the amazing software that gives the Linotronic imagesetter its ability to produce desktop published documents, software known as PostScript.

# Introduction to PostScript

PostScript is a page description language. In many respects, it is like other computer languages such as Basic or Pascal. But unlike other computer languages, PostScript does not run on your computer. Rather, it runs on an output device such as a laser printer or imagesetter.

### How PostScript Works

It is not necessary to understand how PostScript works in order to use its power and flexibility. You need only use your publishing software the way you normally do to access the full range of PostScript functions. Nonetheless, it is worthwhile to have a limited understanding of how the page description language works.

The most important thing to understand about PostScript is that it is a programming language with all the power and flexibility of languages that run on computers. Unlike general-purpose computer languages, PostScript has one main function: to produce images in the form of dots on an output device. It uses English-like commands that can create almost any kind of graphic image or text a user might desire.

Because it is a programming language, a user can theoretically write a PostScript program to produce pages. If we had wanted, we could have written this entire book as a PostScript program. However, this is not very practical; writing PostScript programs requires considerable training and skill. However, most desktop publishing and graphics programs have the ability to generate PostScript files. Sometimes this is handled by the program itself. In other cases, it is handled by the general system software.

When you print a file on a PostScript laser printer or imagesetter, your software generates a PostScript pro-

**Figure 2-5.** Device independence is one of PostScript's greatest advantages. The same PostScript file that produces a page at 300 dpi on a laser printer (top) can produce the page at up to 3386 dpi on an imagesetter (bottom).

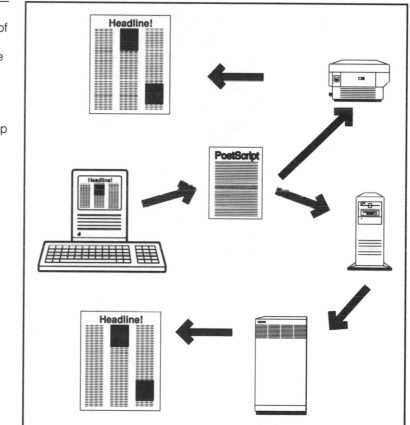

gram and sends it to the output device. There, a controller or RIP converts the commands in the PostScript program into instructions that guide the printer as it produces the page (Figure 2-5). If you were to print the output to a disk file, you would be able to open it with a word processor and see the PostScript commands used to produce the page (Figure 2-6).

**Figure 2-6.** A PostScript program uses a series of English-like commands to describe how an image shold be produced. This program generates a box.

```
%!PS-Adobe-2.0 EPSF-1.2
%%Creator: Adobe Illustrator(TM) 3.0b1r1
%%For: (Steve) (MPR)
%%Title: (ch2-7.ps)
%%CreationDate: (9/7/90) (4:49 PM)
%%DocumentProcSets: Adobe_Illustrator_1.1 0 0
%%ColorUsage: Black&White
%%DocumentProcessColors: Black
%%BoundingBox:59 375 222 492
%%TemplateBox:288 360 288 360
%%TileBox:0 0 552 730
%%DocumentPreview: None
%%EndComments
%%EndProlog
%%BeginSetup
%%EndSetup
u
0.5 g
0 G
0 i 0 J 0 j 1 w 4 M []0 d
%%Note:
220 377 m
220 490 L
61 490 L
61 377 L
220 377 L
b
140.5 433.5 m
B
U
%%Trailer
```

## Benefits to the Desktop Publisher

PostScript is particularly worthwhile for desktop and professional publishing users. There are many reasons for this, including:

- Font quality
- Typeface variety
- Picture quality
- Device independence
- Support for publishing software

Let's consider each of these advantages in more detail.

### Font Quality

PostScript users have access to the finest quality fonts available for their machines. These include fonts from the Linotype Typeface Library licensed by Linotype. While the actual quality of type produced on a printer or imagesetter depends heavily on the resolution of that device, PostScript ensures that font quality will be the best possible at that resolution. For example, if your laser printer has a resolution of 300 dpi, PostScript cannot image fonts at a resolution greater than 300 dpi. What it can do is optimize the placement and arrangement of dots on the page so that each character in each font is true to the original typeface design and as readable as possible.

### Typeface Variety

Another great benefit is the wide variety of typefaces and point sizes available to PostScript users. Unlike other laser printers and typesetters, PostScript devices need store only one representation of each typeface. Any point size can be derived from this compact and precise mathematical representation. This saves the user and output device a tremendous amount of memory compared to the devices that force you to store, and many times purchase, a separate version of each typeface in each point size.

PostScript users can also select from a vast array of typeface families. Many companies, including Linotype and Adobe Systems, provide typefaces in PostScript formats. The number of available fonts is increasing steadily.

### Device Independence

PostScript users also enjoy the benefit of device independence. This means that a page or publication produced with a program that supports PostScript can be generated with little or no modification on any PostScript printer or imagesetter. No matter what the resolution of that output device, the page can be produced at the maximum resolution and with the maximum quality.

### Picture Quality

PostScript output devices possess a unique and impressive capability to produce halftones, a type of photograph that has been converted into a form that can be reproduced on a printing press. While other types of devices can generate such images, PostScript printers and imagesetters are equipped with a vast array of image manipulation tools. As we will see in Chapter Six, PostScript can control the line screen, number of gray levels, contrast, shape, orientation, clipping region, and a host of other attributes affecting a halftone's appearance. The language also includes many functions for producing special effects with text and non-halftone graphics.

### Support for Publishing Software

In the last few years, PostScript has become the most important output standard for developers of high-quality graphics and publishing software. Desktop publishing and computer graphics users have immediate access to a wide variety of programs running on many different computers that support PostScript. As a user, you can be assured not only that a wide variety of programs will be available, but more important, that those programs will perform at their best when a PostScript output device is used. Most of this software can support non-PostScript printers, but in many cases the program's full range of features will not be available to users without PostScript devices.

Later in this book, we will cover many specific characteristics of PostScript, including its font-handling functions and ability to produce gray-scale and color images. But first we need to discuss other hardware and software that will help you get the most out of your Linotronic imagesetter.

# CHAPTER 3

# Front-End Environments

The Linotronic imagesetter represents one of the final steps in the process of producing a publication. For most users, this process begins with a computer system equipped with publishing, graphics, and word processing software. These systems, known as "front ends," are the user's principal means of interacting with the imagesetter. Here is where articles are composed, images rendered, and text and graphics integrated into page layouts.

Most front ends for PostScript-compatible Linotronic imagesetters are Macintosh or IBM-compatible microcomputers. These environments have the two essential elements for communicating with a Linotronic imagesetter: a compatible hardware connection and the ability to produce PostScript output. Both environments also benefit from a wide range of software that can take advantage of PostScript's power to produce quality text and graphics.

In this chapter, we will explore the front-end environments most often used with PostScript-based Linotronic imagesetters. We'll begin by covering the two principal hardware platforms, Apple's Macintosh and the IBM compatibles. In our discussion of IBM compatibles, we'll look at graphical operating environments like Microsoft Windows and GEM that offer many of the user-friendly advantages of the Apple machine. The chapter concludes with a discussion of software packages frequently used with Linotronic imagesetters, including graphics and desktop publishing programs.

The Linotronic imagesetter is a "multilingual" device. In addition to understanding PostScript, Linotronic imagesetters can be set up to work with front-end systems that use typesetting languages like CORA and DENSY. However, the focus of this book is on PostScript-based imagesetters used with standard desktop publishing and computer graphics software. Some of the alternative front-end systems run on standard Macintosh or IBM-compatible microcomputers, but they require that you learn complex code-based typesetting commands. They also lack the graphics power of PostScript. For these reasons, we will limit our discussion to front-end systems that produce PostScript output.

# Apple Macintosh

Apple's Macintosh computer, introduced in 1984, brought several innovations to the mass microcomputer market. Based on an earlier, ill-fated product known as the Lisa, the Macintosh had a built-in "graphical user interface" designed to simplify its use. This interface uses small pictures known as "icons" to represent files and common functions. The Macintosh also includes a device called a "mouse" that allows users to point at the icons on the screen. Moving the mouse along the desktop causes a corresponding movement of the cursor on the screen. Instead of having to memorize a series of commands for copying, moving, or deleting files, users merely point at the file's icon with the mouse. To delete a file, for example, you just drag it to the icon of a trash can.

The graphical user interface was an important innovation. But the Macintosh was also the first operating environment to support PostScript. Apple's LaserWriter printer, introduced in 1985, was one of the first PostScript printers

**Figure 3-1.** Apple's Macintosh computer and LaserWriter printer helped spur the popularity of PostScript.

available. Its popularity among desktop publishing users did much to advance the acceptance of PostScript as a standard (Figure 3-1). The combination of PostScript and the computer's built-in graphics capabilities made the Macintosh a natural choice for publishing applications, and it remains popular among desktop publishing and computer graphics users.

## Connecting to the Macintosh

Any model of the Macintosh can produce output on a Linotronic imagesetter as long as the computer is equipped with an AppleTalk port. This connector is both physically and electronically distinct from the serial or parallel interface ports used by other types of computer printers. The AppleTalk connection allows the Macintosh to communicate with PostScript printers at a rate of 230,000 bits per second—more than 20 times faster than the serial communication used by other computers. It also allows the imagesetter to operate on the AppleTalk network, so that many users can share the printer. To connect your PostScript printer to the AppleTalk network, you must purchase an AppleTalk connector kit for the printer and for each Macintosh you would like to use on the network.

Keep in mind that if you use a service bureau's imagesetter, the issue of connecting your Macintosh is irrelevant. Instead of printing directly to the imagesetter, you will copy your files to disk or transmit them to the service bureau over a modem. However, you will most likely need an AppleTalk connection if you want to produce page proofs on your PostScript laser printer.

### Software Support

Apple Macintosh users enjoy a special advantage with respect to PostScript: nearly every software application that runs on the Macintosh computer can produce PostScript output. The reason for this is that Apple engineers built PostScript support into its standard printer "driver," a term that refers to the software that manages printing operations (Figure 3-2). Any application that fully conforms to the Macintosh operating system automatically supports PostScript. There is no need for the software developer to write a special PostScript driver.

The standard Macintosh driver supports several key areas of PostScript output, including:

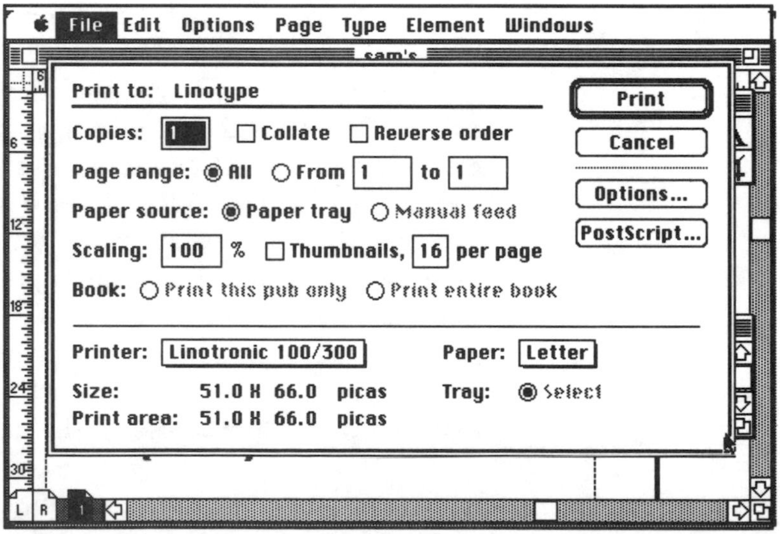

**Figure 3-2.** The Macintosh system includes a built-in printer driver for PostScript.

**FRONT-END ENVIRONMENTS**

- Fonts
- Object graphics
- Bit-mapped graphics
- Network performance

*Fonts*

The Macintosh supports an unlimited number of fonts and a nearly infinite selection of point sizes for each font. Once you have installed a font in the system file of your Macintosh, all of your applications will have access to it (Figure 3-3). In most applications, you select your choice of fonts, type style, and point size from menu options labeled Font, Format, or Style. The many issues involved in working with fonts are explained in greater detail in Chapter Four.

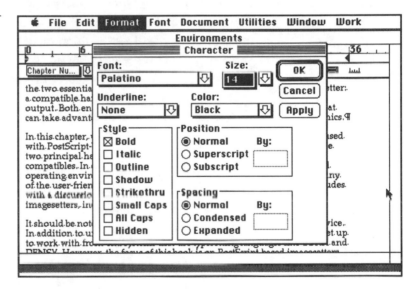

**Figure 3-3.** Once you install a font in the Macintosh system, any application has access to it.

*Object Graphics*

Computer systems generally support two types of graphic images, object graphics and bit-mapped graphics. Object graphics are elements such as circles, lines, rectangles, and shading patterns produced by drawing programs such as MacDraw or SuperPaint, which are discussed in greater detail later in this chapter. When sent to the printer, each graphic element is converted from its internal Macintosh representation into the appropriate PostScript language command. This ensures that object graphics will be imaged at the maximum resolution of the PostScript printer or imagesetter. Thus, circles or rectangles that appear jagged on the Macintosh display will appear smooth on output.

Object graphics can also be enlarged or reduced without loss of quality. For example, if you import a drawing produced in MacDraw into PageMaker and then enlarge it by a factor of two, the image will not lose any quality when sent to your PostScript printer.

### Bit-Mapped Graphics

Bit-mapped graphics are the paint-type images created with such programs as MacPaint or FullPaint, or captured by means of a scanner. These images are sent on a dot-for-dot basis from the Macintosh to your PostScript printer. As a result, they will not appear any better in printed form than they did on screen. Also, if you enlarge a bit-mapped image at output time, the effective resolution of the printed image will be diminished. These types of images are discussed in greater detail later in this chapter.

### Network Performance

The print manager software built into the Macintosh system not only handles the font and graphic conversions discussed above, it also manages the details of getting pages from the Macintosh to the PostScript output device. In a network environment, with many users seeking access to the printer at the same time, this is not a straightforward task. Each time a Macintosh user tries to print, the computer transmits a message over the AppleTalk network to see if the chosen output device is available. If so, it sends the PostScript commands that represent each page of output over the AppleTalk network to the printer. If the printer is not available or in use by another network user, the Macintosh sends an appropriate message to the user.

If a print spooling program like Apple's PrintMonitor is

installed, you needn't wait for the computer to send all of the PostScript commands to the printer. You also needn't worry about someone else using the printer at the same time. Instead of sending output directly to the printer, the Macintosh will send it to a disk file. This disk file will be printed when the output device becomes available; the computer can still be used while printing takes place.

## PostScript Printing Options

PostScript printers offer several unique options for Macintosh users. At print time, you can elect to enlarge or reduce your printed pages by a specified percentage. PostScript handles all the details of producing text and graphic images in the proper sizes without compromising image quality. You can also choose to invert your output so that a mirror image is produced. This is useful for certain photographic options in lithography. Similarly, you may choose to produce a photographic negative of your pages, so that black becomes white and vice versa. Again, this is useful for producing images directly on film, a process that can save you the cost of camera work during offset reproduction.

## Applications that Support PostScript Directly

The capabilities described above are available to any Macintosh user with the standard built-in printer driver. As powerful as this driver is, it does not tap the full potential of the PostScript language. For this reason, a growing number of Macintosh applications support PostScript directly, and offer graphic effects beyond the capabilities of Apple's print driver. These effects are still available only if you have a PostScript output device. They include:

*Text Rotation*
PostScript can rotate text by any number of degrees.

*Gray-Scale Images*
As we will see in Chapter Six, PostScript has very powerful

capabilities for imaging professional-quality halftone photographs. At present, the standard Macintosh printer driver does not make full use of this feature.

### Color Images

Similarly, PostScript has a robust set of features for handling color images. Currently, the Macintosh print driver cannot directly output color PostScript images.

### Fills and Fountains

Graphic programs such as Adobe Illustrator are able to produce stunning special effects such as graduated fill patterns and starbursts. These special effects generally cannot be accessed through the Macintosh printer driver (Figure 3-4).

**Figure 3-4.** Some PostScript features, such as the ability to produce these fountain effects, are not available through the standard Macintosh printer driver.

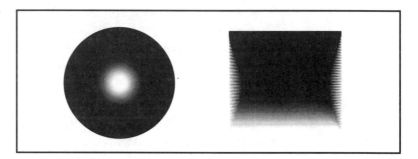

# IBM-Compatible Environments

The IBM PC and compatible computers represent the most popular computer environment available today (Figure 3-5). Unlike the Macintosh, where a single company—Apple Computer—is responsible for manufacturing the computer and developing the system software, the IBM environment is a free-for-all where no single company has absolute control. IBM, which developed the original IBM Personal Computer with its Disk Operating System (DOS), no longer manufactures that particular computer. How-

**Figure 3-5.** IBM-
compatible computer
system.

ever, many other companies make computers compatible
with the PC and its successors, the XT and AT. IBM offers
a line of microcomputers known collectively as the Personal
System/2, which are in many ways incompatible with the
older machines.

The most influential company in terms of IBM-compatible
software is Microsoft, which developed the DOS operating
system and is also a major developer of applications software. Still, the IBM-compatible environment lacks the
kind of conformity made possible in the Macintosh environment where a single company has control. This situation,
as we shall see, has benefits and drawbacks.

IBM-compatible computers can connect with Linotronic
imagesetters through one of four interface ports: serial,
parallel, AppleTalk, or the optional Ethernet interface.
AppleTalk or Ethernet connections are used when the
computer is part of a local-area network; they represent the
fastest form of communication, but are also the most
expensive because you must install a network interface
card in the computer. The serial or parallel interfaces are
used when a single computer is connected to the imagesetter.
However, serial connections tend to be very slow.

To take full advantage of the graphics capabilities in PostScript, an IBM-compatible computer must have a bit-mapped display. In a bit-mapped display system, the software controls each dot on the screen individually, allowing the computer to show graphic images in addition to text. The Macintosh includes a built-in bit-mapped display with 72 dpi. The standard DOS display, on the other hand, is limited to showing characters in an 80-column by 25-line grid. But through a standard graphics display adapter, such as Hercules, CGA, EGA, or VGA, you can add bit-mapped display capabilities to an IBM-compatible computer. Almost all monochrome monitors sold with IBM compatibles include a Hercules-compatible adapter. Most color monitors are sold with a VGA adapter.

One problem with IBM-compatible computers is that many people find the DOS operating system difficult to use. To master DOS, you need to learn a series of commands that are entered from the keyboard. For this reason, some software developers have designed "graphical user interfaces" that provide PC users with environments offering many of the advantages of the Macintosh. Like the Macintosh with its standard printer driver, the DOS-based GUIs provide built-in drivers that support PostScript output. Software written to run under these interfaces can provide the same kind of access to PostScript's powerful graphics features found in the Macintosh. The difference is that much of the work of implementing this PostScript support is left to the individual software developer rather than being handled system-wide.

The most widely used GUI is Microsoft Windows. Others include GEM, from Digital Research, and the OS/2 Presentation Manager, also from Microsoft. In general, these environments work only with software packages that have been written specifically for them. For example, you cannot run the Windows version of Ventura Publisher under the GEM interface.

**Figure 3-6.** Microsoft Windows is a graphical user interface for IBM-compatible computers.

In addition to these standard operating environments, many developers offer stand-alone applications that can produce PostScript output. Some of these applications run under their own graphical environments, while others, primarily word processing programs, run under the DOS character-based interface.

## Microsoft Windows

Microsoft Windows (Figure 3-6) is the most popular graphical operating environment for DOS computers. The early versions of the system suffered from slow performance and memory management problems, but these have been addressed with later versions of the environment beginning with Windows version 3, which was introduced in 1990. Many developers of desktop publishing and graphics programs for the Macintosh offer versions of their packages for Windows.

Microsoft Windows includes a driver for PostScript output (Figure 3-7). As a result, nearly any program that works with Windows can produce output on the Linotronic imagesetter. The Windows Control Panel also offers func-

**Figure 3-7.** Windows provides a built-in printer driver for PostScript.

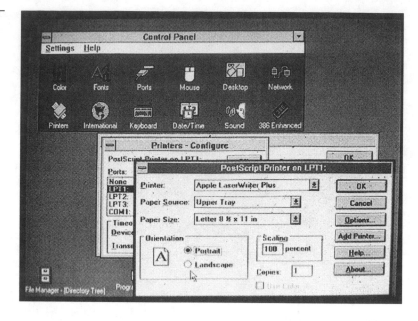

tions that allow you to choose and configure the printer port with which the computer sends data to the imagesetter. In many cases, users will choose to use a LAN compatible with AppleTalk or Ethernet to communicate with the imagesetter.

Windows also offers functions that allow the user to install screen and printer fonts. Printer fonts are automatically loaded when the user installs the output device. As with the Macintosh, users can install an unlimited number of fonts and point sizes. Again, this is handled through the Windows Control Panel. Once the font is installed, it is available to all Windows applications. However, you can also install fonts that are application specific and work only with a particular program, such as the Corel Draw illustration package.

Like the Macintosh, Windows supports a wide range of capabilities in graphics and desktop publishing applications, including object-oriented and bit-mapped images. However, the operating system places more responsibility on individual applications to perform many software operations that are handled system-wide on the Macintosh.

This reliance on individual applications also carries over to network software. On installation, Windows can automatically detect the presence of a local-area network. However, it does not offer built-in functions for handling user conflicts and other networking issues that are handled by the Macintosh system software. Instead, these tasks are performed by the specific network operating system installed alongside Windows.

### GEM

The GEM environment (Figure 3-8), from Digital Research Inc., is one of the oldest graphical environments for IBM-compatible computers. One of the most popular programs that runs under GEM is Xerox Ventura Publisher. The advantage of GEM is that users of less powerful machines such as the IBM XT and compatibles can make use of a graphical environment without a severe performance penalty. Besides Ventura Publisher, other popular programs that run under GEM include Artline, an illustration program from Digital Research, and Per:Form, a forms design program from Delrina Technology.

Like Windows, GEM includes a PostScript printer driver

**Figure 3-8.** Many IBM PC programs, such as this version of Ventura Publisher, run under the GEM environment.

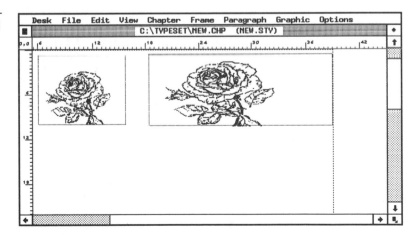

that is available to any application that runs under the environment. However, some GEM programs, such as Ventura Publisher, incorporate their own PostScript drivers that enhance the quality of Linotronic imagesetter output.

## OS/2 Presentation Manager

OS/2 is the next-generation operating system announced by IBM in 1988 and manufactured by Microsoft. The Presentation Manager is a version of OS/2 that incorporates a GUI very similar to that in Microsoft Windows. OS/2 requires more memory than Windows, but it has many new capabilities of interest to desktop publishing and graphics users. Among these are the ability to run several applications simultaneously, and to have those applications share information among themselves while they are running.

## Stand-Alone Applications

Aside from the three graphical operating environments we discussed above, many stand-alone applications for the IBM PC and compatibles can also produce output on Linotronic imagesetters. Most word processing programs that run under a character-based interface on DOS offer PostScript output as an option. The problem with many of these programs is that you don't get a good idea of how your pages will look until they are printed. In many cases, files created by a word processing program will be imported into page layouts created with a desktop publishing package that runs under GEM or Windows.

In addition to character-based programs, some PostScript-compatible software packages run under their own proprietary graphical environments. These include PFS:First Publisher, an entry-level desktop publishing program from Software Publishing Corp., and Gray/FX (Figure 3-9), a halftone image-editing program from Xerox Imaging Systems. We will discuss halftone image-editing software in more detail in Chapter Six.

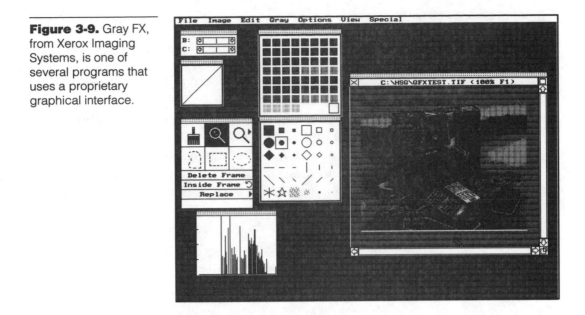

**Figure 3-9.** Gray FX, from Xerox Imaging Systems, is one of several programs that uses a proprietary graphical interface.

# Unix Workstation Environments

Most of this book focuses on the use of Linotronic imagesetters with Macintosh and IBM-compatible computers. But many Unix-based workstations also support PostScript and Linotronic imaging. Unix was originally developed as a science-oriented operating system for university and engineering environments, but it is steadily building up a core of mainstream business applications.

Manufacturers of Unix-based workstations include Sun Microsystems, Hewlett-Packard, and NeXT Computers. The two most popular Unix-based publishing packages are Interleaf Publisher and FrameMaker. These are high-performance programs oriented toward high-volume publishers of books, technical manuals, and proposals. Both can produce PostScript output on Linotronic imagesetters.

# Application Software

So far, we have focused our discussion on computer hardware and system software. But it is the application software—word processing, graphics, desktop publishing, and so on—that makes a computer system truly useful. Almost any kind of program, including spreadsheets, databases, and statistical analysis software, can play a part in the process that ends with high-resolution output on a Linotronic imagesetter. But certain categories of software play an especially important role in producing Linotronic output. The three most important categories are bit-mapped graphics software, object-oriented graphics software, and desktop publishing software.

## Bit-Mapped Graphics

As mentioned above, graphic images produced on a computer system fall into two general categories: bit-mapped and object-oriented (Figure 3-10). Bit-mapped images, simply defined, are images produced as a pattern of dots. They are sometimes described as "pixel-oriented," "raster," or "paint" images. Images displayed on a computer screen are bit mapped. The Macintosh display, for example, shows images at a resolution of 72 dpi. Laser printers and imagesetters also produce bit-mapped images, laser print-

**Figure 3-10.** Object-oriented graphics can be enlarged without loss of image quality. Here, the panda's nose was enlarged as an object-oriented image (middle) and as a bit map (right).

ers at 300 dpi and imagesetters at resolutions up to 3386 dpi. Obviously, at that resolution it is impossible to see the individual dots, but the images are bit mapped all the same.

Bit-mapped graphics programs come in a many varieties. Simple paint programs (Figure 3-11) allow you to create pictures, or retouch scanned images in black and white or a limited number of colors. They typically include a paint-brush, eraser, paint bucket, and a selection of different patterns, colors, and line widths. Most include a zoom-in mode that allows you to edit individual pixels. They are generally easy to use, but they are also limited in the kinds of images they can produce. One limitation is resolution. Some can produce images with up to 300 dpi, but others are limited to 72 dpi.

Image-editing programs (Figure 3-12) are designed to work with black-and-white photographs and other images with multiple shades of gray. At first glance, they seem much like paint packages, with many of the same tools and other similar features. However, they also include features that allow you to manipulate the gray value of each pixel in an image. Some can produce photographic effects previ-

**Figure 3-11.** Paint programs, such as SuperPaint, produce images in the form of dots.

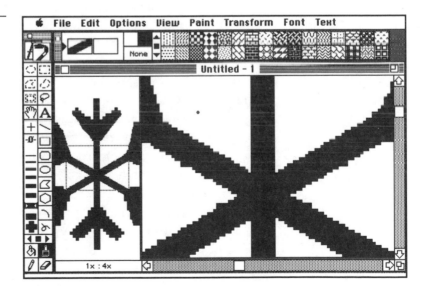

**Figure 3-12.** Image-editing programs are designed to work with photographic images.

ously achieved only in a darkroom. We will describe these packages more extensively in Chapter Six.

Color image-editing programs take image editing a step further and allow you to work with color photographs and other images in which the intensity of individual colors can vary. They tend to be quite sophisticated, and can be used to produce visual effects that once were possible only on expensive color workstations. These programs will be discussed more extensively in Chapter Seven.

## Illustration Programs

The second kind of image is known as a "draw" or "object-oriented" graphic, and it is generally created by a category of software known as *draw* or *illustration* programs (Figure 3-13). True to their name, object-oriented graphics programs treat images as collections of discrete objects. The basic units of these objects are lines and curves. Once lines and curves are used to create an object, the object can be combined and arranged with other objects to create complex illustrations. You can also fill objects with specified colors or shades of gray.

**Figure 3-13.**

Illustration programs use object-oriented graphics tools to produce precise artwork that can be printed at high resolutions.

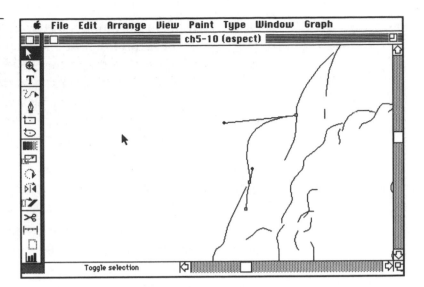

Objects in a paint program are rendered as a pattern of bits, but objects in a draw program are described mathematically. To produce a circle, for example, a draw program uses a mathematical expression that translates into a circle on the screen, and later on the page. Because the image is not locked into a particular bit pattern, it can be printed at the full resolution of the output device, whether a 300-dpi laser printer or 2540-dpi imagesetter. We mentioned earlier that monitors and output devices produce images as patterns of bits. This is still true of object-oriented images. The difference is that the software internally recognizes the objects as mathematical expressions. It is only when the image is displayed or printed that it is converted to the bit-mapped format.

Draw programs allow artists to create complex illustrations with great precision. Because the software recognizes lines and curves as distinct objects, they can be stretched, moved, rotated, or manipulated in other ways. Small objects can be grouped into larger objects, which can also be manipulated.

Most programs in this category use what are known as Bezier curves. Such curves can be moved, resized, and reshaped by means of small handles that appear on or near the curve. These handles are known as control points. Bezier curves permit a great deal of flexibility and control over the slope and length of curves, but using them effectively to create objects takes practice. Most illustration programs have a freehand feature that allows you to draw a curve, then inserts endpoints and anchor points that can be further manipulated.

### Desktop Publishing Packages

For most users, desktop publishing packages (Figure 3-14) are the principal means of producing output on a Linotronic imagesetter. These programs, also known as page layout or page composition packages, allow users to create a wide range of publications using desktop computers. Typically, the user is shown an on-screen representation of a page layout on which text and graphics can be positioned. Most desktop publishing programs have limited tools for creating text and graphics. Instead, they are designed to incorporate text and graphics created with other programs. For example, a user might use a word processing package like Microsoft Word to write articles, a paint program like SuperPaint to create bit-mapped images, and an illustration program like Aldus Freehand to produce object-ori-

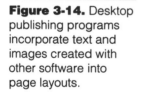

**Figure 3-14.** Desktop publishing programs incorporate text and images created with other software into page layouts.

ented images. These elements are then imported into the desktop publishing program and positioned on the page.

Desktop publishing programs vary widely in their functions and capabilities. Low-end packages are suited mostly for producing flyers and other simple one-page documents. More complex publishing packages include sophisticated features like style sheets, indexing, and typographic functions that offer precise control over character spacing. Some word processing packages have page layout and graphics import features that approach the functionality of desktop publishing software. At the same time, many desktop publishing programs are adding features like spell checking and find-and-replace that were once the province of word processing packages.

Almost all desktop publishing packages are capable of producing PostScript output. In fact, you can write an article or create a graphic image in a program that does not support PostScript, then import it into a publishing package for output on a PostScript laser printer or imagesetter. Some of these programs use the standard PostScript driver provided with the operating environment. Others include their own PostScript drivers that add capabilities not found in the standard driver.

The number of applications and operating environments that support PostScript is growing rapidly—much more so than we could possibly cover in detail in this book. Instead, we will try to acquaint you with general classes of applications. In the next three chapters, we will discuss three important applications relating to Linotronic imagesetters: fonts, halftone images, and color images.

# CHAPTER 4

# Working with Fonts

To live in modern civilization is to live in a world of type. We see it every day, not only in books and newspapers, but also on billboards, cereal packages, street signs and almost any other medium that uses the printed word. It is such an integral part of our lives that many of us ignore it.

Suppose you are driving along and see a sign that says:

**SPEED LIMIT**
**35 MILES PER HOUR**

You might slow down, but would you stop to wonder, "Did they set that in Helvetica Black?" Probably not. But then again, you might give the question some thought if you are one of the growing number of people who use desktop publishing systems.

With the desktop publishing explosion, millions of computer users have learned to appreciate the complexities of type. Thousands of typeface designs have been created for every possible need, from the flashy display fonts used in print advertising to the subtler text faces seen in newspapers, magazines and books. But to employ these typefaces effectively, users must deal with such mundane technical matters as "font downloading," "screen fonts" and "font ID numbers." This is especially true when it comes to PostScript imagesetters like the Linotronic series.

In this chapter, we'll explore the many issues involved in

using type with Linotronic series PostScript imagesetters. First we'll discuss terms like "typeface" and "font" and show how their meanings have changed with the advent of desktop publishing. Then we'll explore the characteristics that distinguish one typeface from another. We'll look at the various sources from which typefaces can be obtained. Finally, we'll discuss the technical issues involved in using type in Macintosh and IBM-compatible computer systems.

# Terminology

Desktop publishers hear the terms "font" and "typeface" used interchangeably, but it was not always so. In the pre-PostScript typesetting era, the term "typeface" referred to a specific design of type, such as Times Roman or Helvetica Bold. "Typeface family" referred to a group of typefaces installed as a unit, such as "Times Roman, Times Bold, Times Italic and Times Bold Italic." "Font" referred to a specific typeface in a specific size, such as "14-point Times Bold" (Figure 4-1).

These distinctions made sense at the time. In older typesetting systems, fonts were loaded in distinct sizes and styles. A typesetter equipped with 14-, 18-, 24-, and 36-point Times Bold, for example, would be unable to produce a headline in 30-point Times bold unless you added that font.

In contrast, desktop publishing systems use "scalable fonts." Instead of loading fonts in distinct sizes, you load one copy of each typeface, such as Times and Helvetica in normal, bold, italic, and bold-italic styles. When you specify a type size in a desktop publishing program and print the document, the PostScript software enlarges the typeface to the desired size.

This is why the meaning of "font" has changed. To a PostScript user who can scale type to almost any size, "font" generally refers to a typeface or even an entire typeface

**Figure 4-1.** The term "typeface" traditionally refers to a design of type, such as Helvetica, Helvetica Italic, Helvetica Bold, or Helvetica Bold Italic. The term "typeface family" refers to all of the variations of a type design "Font" refers to a typeface in a particular size, such as 30-point Helvetica Bold. However, many desktop publishers use the terms "typeface" and "font" interchangeably.

Helvetica
**Helvetica Bold**
*Helvetica Italic*
***Helvetica Bold Italic***
30-point Helvetica

family. "Typeface" generally refers to the design of the type. You don't hear people say, "I just installed 14- and 18-point Broadway." Instead, they've installed a copy of the Broadway display face that can be enlarged from four to 1000 points and any size in between (depending on the capabilities of your software).

As we have seen, these sizes are stated in "points," the unit long used in the print production business to measure type size. One point is equal to approximately one-seventy-second of an inch. The "point size" of a font is the maximum amount of space required for characters. For example, if you set the word "High Jump" in 72-point Helvetica, it will measure about 1 inch from the top of the "H" to the bottom of the "p."

## Typeface Characteristics

Typeface design is an age-old art form that dates to the times of Gutenberg. Type designers say that creating a new typeface is an artistic achievement on the level of a novel or a symphony. Each typeface has unique characteristics, and usually represents many long hours of design work even with the computer-aided design software available today.

Typographers use a rather large vocabulary to describe the shapes of characters (Figure 4-2). An ascender is the portion of certain lowercase characters—b, d, f, h, k, l, t— that rises above the main body of the character. A descender is the portion of other lower-case characters—g, j, p, q, y—that descend below the baseline. "X-height" is the main body of a character measured from the baseline, without ascenders or descenders. If you look through a type catalog, you'll notice that some fonts have larger x-heights than others. In addition to ascenders and descenders, type designers use terms like "stem," "link," and "bowl" to describes the sections of a character shape.

**Figure 4-2.** Type designers use many terms to refer to the parts of a character.

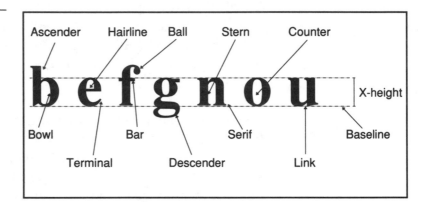

## Serifs

One feature that immediately distinguishes one font from another is the presence or absence of serifs—tiny crossbars used on the ends of character shapes (Figure 4-3). Typefaces with serifs are often used for body text in newspapers, books, and magazines because many people regard them as being easy to read. Typefaces without serifs—sans serif typefaces—are often used in headlines. The most well known serif typeface is probably "Times," while the most well known sans serif typeface is "Helvetica."

**Figure 4-3.** Serif type
(top) and sans-serif
type (bottom).

The quick brown fox jumps over the lazy dog.
The quick brown fox jumps over the lazy dog.
The quick brown fox jumps over the lazy dog.
The quick brown fox jumps over the lazy dog.

The way serifs are designed in a particular typeface often provide its special look (Figure 4-4). Optima, for example, has small, barely noticeable serifs. Garamond has rounded, triangular serifs. Lubalin Graph has large, squared-off serifs. Other features that distinguish one typeface from another include its character width, the amount of difference between thick and thin lines, and a factor known as stress, which refers to the degree of slant in the main body of the character.

**Figure 4-4.** Serifs can
give a typeface much
of its unique look.

**Bodoni**
Garamond
Lubalin Graph
Melior
Optima

Linotype has developed a classification system for typeface designs based on factors like stress and the use of serifs (Figure 4-5). "Old Face" fonts, such as Berkeley Oldstyle, are characterized by light weights and small, bracketed serifs. "Transitional" faces like Bookman have stronger

Old Face: ITC Berkeley Oldstyle
Transitional: Bookman
Modern: New Century Schoolbook
Slab serif: Lubalin Graph
Sans serif: Helvetica
**Decorative: Bodoni Poster**
*Script: Zapf Chancery*
Blackletter: Linotext

serifs and greater contrast between thick and thin lines. "Modern" faces, such as New Century Schoolbook, have fine serifs, a vertical stress, and even more contrast between thick and thin lines. "Slab serifs" like Lubalin Graph have heavy rectangular serifs with little difference between thick and thin lines. The "sans serif" category applies to all faces without serifs. "Decorative" typefaces generally feature unusual designs meant for display purposes. "Script" or "brush" faces present a handwritten or calligraphic image. "Blackletter" typefaces, such as Linotext, are most often seen in the banners of the *New York Times* and other newspapers. These categories are not all-encompassing, and some typefaces have features that straddle multiple design classifications.

One advantage of the Linotronic imagesetter is its ability to handle a wide range of font designs. One example is Bodoni, the typeface used for body text in *The New Yorker* magazine. With its hairline serifs and thin horizontal strokes, Bodoni does not reproduce well at small point sizes on 300-dpi laser printers. But the Linotronic imagesetter provides faithful reproduction of this difficult typeface even at 1270 dpi.

Some fonts are better suited for certain kinds of documents than others. For example, Zapf Chancery would probably look rather silly in a hardware store ad. And Cooper Black would definitely be out of place on a wedding invitation. Certain combinations of headline and text fonts also look better than others. Many books on graphic design provide guidance on these subtleties of font usage.

## Font Variations

Most typefaces are created in four to six variations. The "plain" or "roman" style is the basic version of the typeface intended for use as body text. "Italic" refers to a slanted design used in body text for titles and emphasis. Sometimes the term "oblique" is used to describe roman character designs that are slanted by the software rather than being designed from scratch. "Bold" refers to a thick, heavy version of the typeface, used in headlines and also for emphasis in body text. "Bold italic" combines the bold and italic styles. Some typefaces also come in condensed and expanded variations. Others come in a "light" version slightly thinner than the roman weight, or a "demibold" version slightly thicker than roman but thinner than bold. Some typefaces can be printed in "outline" or "shadow" formats. Instead of being installed as separate fonts, these styles are created from the outlines of existing fonts. Some decorative and script typefaces, such as Cooper Black, Dom Casual, and Park Avenue, are available in one or two weights only.

Another factor that distinguishes typefaces is proportional versus monospaced character spacing (Figure 4-6). In most typefaces, characters occupy varying amounts of space depending on their size. An "I," for example, occupies less space than an "M." This is known as "proportional spacing." In a monospaced typeface, each character occupies a fixed amount of space no matter what its size. Proportional typefaces are generally preferred because they are easy to read and economize on space. But monospaced fonts are sometimes used for certain effects, such as imitating the

**Figure 4-6.**
Proportional spaced
type (top) uses variable
spacing based on
character widths.
Monospaced type
(bottom) uses a fixed
amount of space for
each character.

This is an example of a proportional spaced font.

This is an example of a monospaced font.

output of a typewriter. The popular Courier typeface, for example, is a monospaced font designed for this very purpose.

## Character Sets

One more distinction among fonts is the character sets they include. Almost every font includes the standard alphabet in upper- and lower-case plus numerals and the symbols commonly found on a typewriter. Most typefaces also include additional symbols found in what's known as the standard ASCII character set. To produce one of these symbols, you type an ASCII code or a combination of keys, depending on your hardware and software. For example, you can enter a bullet symbol (•) on a Macintosh by typing an "8" while you hold down the Option key.

Some specialty typefaces include non-English language character sets or symbol collections. One of the most popular symbol typefaces is Zapf Dingbats, which includes snowflakes, check marks and other symbols. Each symbol is linked to a standard key on the keyboard. When you type an "A" in Zapf Dingbats, you get a Star of David (✡).

# Typeface Ownership

As we have seen, typefaces are referred to by specific names —Times Roman, Helvetica, Palatino and so on. Many of these names are owned by typesetter manufacturers and other companies as trademarks for their typefaces. Times

Roman, Helvetica, and Palatino, for example, are owned by Linotype AG and/or its subsidiaries. This situation dates back to the days before desktop publishing when typographers purchased their equipment and fonts from the same source. If you bought a Linotype typesetter, for example, you also had access to Linotype fonts. If you bought a competing typesetter, you had access to that manufacturer's fonts. Some typefaces are part of the public domain and can be used by any manufacturer. Other typefaces are owned by organizations like the International Typeface Corp. (ITC) that offer them to any manufacturer willing to pay the price.

But what if you were a non-Linotype manufacturer who wanted to offer a Linotype typeface? Trademark laws protect the name of a typeface. However, type designers are free to create typefaces that look like existing designs as long as they use a different name. Compugraphic's "Helios" and "Palacio," for example, are redesigned versions of Linotype's Helvetica and Palatino. This practice continues today among vendors of fonts for laser printers and imagesetters. However, if you want the original design of a licensed typeface, you must use the version from the company that owns the license.

The growth of desktop publishing has created a revolution in the way typefaces are sold. One integral part of a desktop publishing system is a laser printer, and sales of laser printers created an enormous demand for fonts. Suddenly, it made sense for companies to sell fonts as commodities rather than offering them exclusively with a particular piece of typesetting equipment. Many of these fonts are designed for use with Hewlett-Packard's LaserJet (and compatible) printers. But many others work with PostScript printers like Apple's LaserWriter II. Given the power of PostScript, these fonts also work with Linotronic PostScript imagesetters.

Many long-established leaders in the typography business

have taken advantage of this new demand for type, including Linotype, Compugraphic, URW and Monotype. Linotype began offering PostScript fonts in August 1987 and its contributions represent about 50 percent of the current market. It offers the fonts under its own label and also through a licensing agreement with PostScript developer Adobe Systems. Adobe has also converted public domain typefaces and licensed designs from other sources such as ITC, and is also one of the leading typeface vendors in the microcomputer market.

In addition to the well-established typesetter manufacturers, many smaller companies have also entered the digital font market. Some limit themselves to producing a small number of original font designs. Others combine original typefaces with alternative versions of popular fonts like Times Roman and Helvetica. One of the best known companies in this category is Bitstream, which was founded in Cambridge, MA., by a number of former Linotype employees. Bitstream bills itself as the first independent "digital type foundry," capable of designing and selling typefaces for a wide variety of computer output devices, including PostScript printers and imagesetters. Bitstream's best-known typefaces are "Dutch" and "Swiss," which are renamed and redesigned versions of Linotype's Times Roman and Helvetica. Its large library also includes original type designs and typefaces licensed from independent type sources like ITC.

# Font Standards

Adobe's success with PostScript is largely due to the inherent power and device-independence of the page description language. But the company also owes much to Apple's use of PostScript in the popular LaserWriter printer. The LaserWriter, one of the first printers to use PostScript, has since been replaced by the LaserWriter II. But it has had an enormous impact on the desktop publishing market, even in the IBM environment, by establishing certain typefaces as font "standards."

The original LaserWriter was sold with 13 resident fonts that have come to be known as the "LaserWriter 13." These include Linotype's Times Roman and Helvetica in normal, bold, italic, and bold italic styles, plus Courier, a public domain typeface once used in many typewriters, in the same four styles. The last of the "LaserWriter 13" is the Symbol font, which includes non-English characters used for special purposes like mathematical formulas.

These 13 fonts are said to be "resident" in the printer. A resident font is sold with your printer or imagesetter and resides there permanently. This means that you can produce any document that includes that typeface without "downloading" the font (copying it to the printer) beforehand. If a font you want to print is not resident, you must download it to the laser printer or imagesetter. As we'll see later, each computer system has its own method for downloading fonts.

The LaserWriter Plus, an upgraded version of the LaserWriter, included 35 resident fonts that are known as the "LaserWriter 35" (Figure 4-7). In addition to the fonts described above, they include Palatino, Avant Garde, Bookman, New Century Schoolbook, and Helvetica Narrow in roman, bold, italic, and bold italic styles, plus two "specialty" fonts, Zapf Chancery and Zapf Dingbats. Zapf Chancery is an ornate, script-like typeface that would be well at home on a wedding invitation. Zapf Dingbats is a collection of symbols, such as snowflakes, check marks and boxes, that help add spice to a document. Both were created by Hermann Zapf, a legendary type designer who also created Linotype's Palatino.

Almost all PostScript printers sold these days include the "LaserWriter 13," and most include the additional 22 fonts in the LaserWriter Plus. The one exception is Helvetica

Avant Garde
abcdefghijklmnopqrstuvwxyz

*Avant Garde Italic*
*abcdefghijklmnopqrstuvwxyz*

**Avant Garde Bold**
**abcdefghijklmnopqrstuvwxyz**

***Avant Garde Bold Italic***
***abcdefghijklmnopqrstuvwxyz***

Bookman
abcdefghijklmnopqrstuvwxyz

*Bookman Italic*
*abcdefghijklmnopqrstuvwxyz*

**Bookman Bold**
**abcdefghijklmnopqrstuvwxyz**

***Bookman Bold Italic***
***abcdefghijklmnopqrstuvwxyz***

Courier
abcdefghijklmnopqrstuvwxyz

*Courier Oblique*
*abcdefghijklmnopqrstuvwxyz*

**Courier Bold**
**abcdefghijklmnopqrstuvwxyz**

***Courier Bold Oblique***
***abcdefghijklmnopqrstuvwxyz***

Helvetica
abcdefghijklmnopqrstuvwxyz

*Helvetica Oblique*
*abcdefghijklmnopqrstuvwxyz*

**Helvetica Bold**
**abcdefghijklmnopqrstuvwxyz**

***Helvetica Bold Oblique***
***abcdefghijklmnopqrstuvwxyz***

Helvetica Narrow
abcdefghijklmnopqrstuvwxyz

*Helvetica Narrow Oblique*
*abcdefghijklmnopqrstuvwxyz*

**Helvetica Narrow Bold**
**abcdefghijklmnopqrstuvwxyz**

***Helvetica Narrow Bold Oblique***
***abcdefghijklmnopqrstuvwxyz***

New Century Schoolbook
abcdefghijklmnopqrstuvwxyz

*New Century Schoolbook Italic*
*abcdefghijklmnopqrstuvwxyz*

**New Century Schoolbook Bold**
**abcdefghijklmnopqrstuvwxyz**

***New Century Schoolbook Bold Italic***
***abcdefghijklmnopqrstuvwxyz***

Palatino
abcdefghijklmnopqrstuvwxyz

*Palatino Italic*
*abcdefghijklmnopqrstuvwxyz*

**Palatino Bold**
**abcdefghijklmnopqrstuvwxyz**

***Palatino Bold Italic***
***abcdefghijklmnopqrstuvwxyz***

Symbol
αβχδεφγηιφκλμνοπθρστυϖωξψζ

Times Roman
abcdefghijklmnopqrstuvwxyz

*Times Italic*
*abcdefghijklmnopqrstuvwxyz*

**Times Bold**
**abcdefghijklmnopqrstuvwxyz**

***Times Bold Italic***
***abcdefghijklmnopqrstuvwxyz***

*Zapf Chancery*
*abcdefghijklmnopqrstuvwxyz*

Zapf Dingbats
✿✪✳✴✵✶✷✸✹✺✦●○■□▢▣□▲▼◆◇▶▍▌▐

**Figure 4-7.** Apple's LaserWriter includes 35 fonts that are known as the "LaserWriter 35."

Narrow, a "slimmed down" version of standard Helvetica. Though offered in many PostScript laser printers, it is not found in the Linotronic imagesetter or other high-resolution output devices, and for a very good reason.

Helvetica Narrow is not a true font in itself, but is generated in a compressed form from the standard Helvetica using a mathematical formula. Because of this, the characters don't look quite as good as they would if the typeface were created from scratch. It is hard to notice the difference on a 300-dpi laser printer, but not on a high-resolution Linotronic imagesetter. If you want to get the Helvetica Narrow effect on a Linotronic imagesetter, you should use a Linotype font called Helvetica Condensed, which was designed from scratch. If you use the Linotronic imagesetter to print a document with text in Helvetica Narrow, the imagesetter will replace the font with a poorly spaced version of Courier.

## Anatomy of a Font

In the old days of hot-metal typesetters, fonts were metal casts forged in a type foundry. By contrast, the digital fonts in today's Linotronic imagesetters are complex creations that require the traditional skills of a type designer along with the technical know-how of a software engineer. Designing a typeface is difficult enough, but then it must be converted into a format that allows its use within a computer system. And it must look good, with proper spacing and character shapes, when produced at any size.

To a PostScript output device like the Linotronic imagesetter, a "font" is a collection of data. One set of data describes the shape of each character in the typeface. Another set describes the amount of space consumed by each character. Yet another set produces slight alterations in character shapes to improve their quality on a 300-dpi laser printer. This information, like all computer data, is stored in files that can be installed from a standard magnetic diskette.

## Outline Fonts

Typefaces used in PostScript imagesetters are known as "outline fonts" because they are stored as mathematical expressions that produce an outline of the character shape. The advantage of an outline font is that it can be scaled to any size and printed at the maximum resolution of the output device. This is why a font that prints on Apple's LaserWriter at 300 dpi also prints on a Linotronic imagesetter at 1270 or 2540 dpi. The alternative to the outline font is the "bit-mapped" font found in some inexpensive laser printers. Bit-mapped fonts are stored in the printer or computer as a series of dots that can be printed in one size only. Because you cannot scale a bit-mapped font, you have to install each distinct type size you think you'll need. This consumes a lot of disk space and limits your flexibility when producing documents with a variety of type sizes. Bit-mapped fonts are also limited to the resolution at which they were created, usually 300 dpi.

Many vendors of outline fonts have their own formats for storing font information. Adobe's, for example, is known as the "Type 1" format. But all fonts include common elements. Two of these common elements are width and kerning tables.

## Font Metrics Files

A width table contains information about the character widths of a font, more technically known as the "font metrics." Even if two fonts appear to have the identical design, variations in metrics can cause them to be spaced differently even when printed at the same size. This is an important factor when evaluating fonts designed for PostScript "clone" printers and imagesetters. Until recently, Adobe Systems would not allow developers of PostScript clones to use its fonts. As a result, Bitstream and other font vendors developed their own versions of Adobe/

Linotype typefaces that matched the metrics of the Adobe products. In many cases, the design of the new typefaces varied in barely noticeable ways from the Adobe originals. But as long as the spacing was consistent and the font looked good, few users complained. Such consistency is essential if you want to use a PostScript clone printer as a proofing device for the Linotronic imagesetter.

## Kerning

A kerning table contains information that improves the spacing between specific pairs of characters. An upper-case "T" and lower-case "a" would appear to be floating separately if printed with their normal spacing. The kerning table tells PostScript to push the two characters together so that their "boundaries" overlap (Figure 4-8). Many desktop publishing programs allow you to further adjust kerning to suit special design needs, but the kerning table provides the information needed for optimal character spacing under normal conditions.

## Hinting

Some type vendors include a series of software commands known as "hints" that are designed to improve type appearance on laser printers with relatively low resolutions. When a 300-dpi printer produces type in small point sizes, there may not be enough dots to reproduce its exact shape. A "hinting" program slightly alters the shape of a character so that it better fits the printer's 300-dpi dot matrix. The quality of a font vendor's hints may be important to laser

**Figure 4-8.** Kerning functions reduce the space between characters to improve their appearance.

printer users, but is of little concern when producing high-resolution Linotronic output.

In addition to supplying fonts for the printer, most vendors also sell "screen fonts" that can be displayed on your computer monitor. Until recently, all microcomputer screen fonts, even those sold with PostScript devices, were bit-mapped. On the Macintosh, for example, you had to install each size and style of screen font that you thought you would need. If you specified a point size that hadn't been stored, the Macintosh software would scale the font as best it could, producing a rough-looking approximation that was often difficult to read. When you printed your document, however, the PostScript-scaled font would look fine.

Things are beginning to change, and we are now seeing the use of outline fonts for the generation of type on the screen. Adobe's Type Manager, for example, produces good-looking type on the Macintosh screen no matter what point size you have chosen (Figure 4-9). Apple's new System 7 software offers similar capabilities on the Macintosh, while Microsoft is also offering scalable fonts for the Windows environment. Many users, however, must still contend with conventional bit-mapped screen fonts.

**Figure 4-9.** Outline screen fonts (top) can be scaled to any size without loss of quality. Bit-mapped screen fonts (bottom) can suffer from jagged edges if the software tries to scale them.

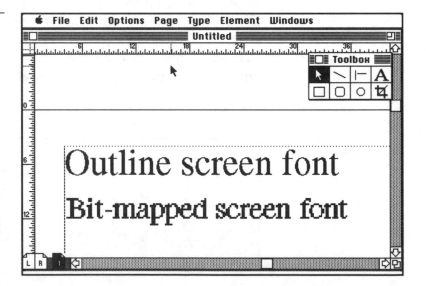

# Font Creation Software

Since digital fonts represent such a complex design task, it might seem that a person would have to be crazy to create one on their own. But several software packages are available for IBM and Macintosh computers that help you do just that.

These packages, which include Altsys' Fontographer, Letraset's FontStudio, and ZSoft's Publisher's Type Foundry, provide numerous tools that allow you to modify existing typefaces or create new ones from scratch (Figure 4-10). Fontographer even includes its own set of hints for optimizing typeface appearance on 300-dpi laser printers. Some professional type designers have used these programs to create commercial font packages. For example, a type designer named Joseph Treacy used Fontographer to create his TF Habitat and TF Forever typefaces for the Macintosh.

Many graphics packages designed for general use include features that allow you to customize headlines created with standard fonts. Corel Systems' Corel Draw, for ex-

**Figure 4-10.**
Fontographer is one of several programs that allow creation of custom typefaces.

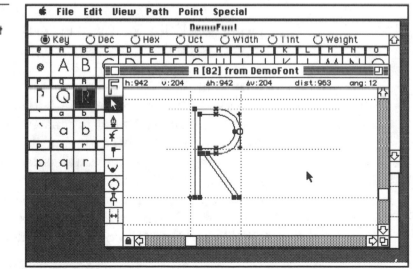

ample, treats PostScript typefaces as graphic objects that can be embellished in many ways (Figure 4-11). Once you have created the desired effect, you can save the headline as a PostScript file and import it into a publishing program.

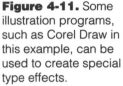

**Figure 4-11.** Some illustration programs, such as Corel Draw in this example, can be used to create special type effects.

# Hardware Environments

Any strategy for using fonts with the Linotronic imagesetter depends in part on the computer environment in which you work. Users of IBM-compatible computers will find that each piece of software has its own approach to font usage. Apple's Macintosh, on the other hand, uses a consistent method for handling fonts across all applications. This distinction is likely to remain even as font technology rapidly advances in both environments.

### Macintosh

Apple's Macintosh represents both good news and bad news for those who want to master the use of fonts. The good news is that most Macintosh applications share a common approach to the use of fonts. The bad news is that understanding this approach can drive the unsuspecting desktop publishing user batty. To learn about Macintosh fonts is to delve into a nether world of suitcases, FONDs, NFNTs and Font ID conflicts. If you don't take the plunge, however, you could find yourself in for an unfortunate surprise when you try to print a document with a fancy-looking typeface on your Linotronic.

As we have noted, certain fonts have become standards on the Macintosh due to their presence in the original LaserWriter printer. These include Times, Helvetica, Courier and Symbol. The fonts are sold with almost every PostScript output device, and should be present in most Macintosh system files. If you produce a document in one of these fonts, you can be sure that it will print on just about any PostScript device without the need to download the fonts. The additional fonts found in the original LaserWriter Plus are also a safe bet, except for the Helvetica Narrow typeface described earlier. If you want to print to a Linotronic in a narrow version of Helvetica, you should purchase the Helvetica Condensed font designed for output on high-resolution devices.

*Third-Party Macintosh Fonts*
Of course, you are not limited to these resident fonts. Hundreds of PostScript fonts are available for all kinds of applications, even bar coding. If you have a laser printer, you will probably have to install any non-resident fonts on your computer's hard disk so they can be downloaded to the printer. Some Linotronic models, however, can store dozens of resident fonts on their own built-in hard disks. Most Linotronic service bureaus offer a wide selection of Adobe/Linotype fonts along with typefaces from other sources, but you'll need to find out how to avoid any surprises when you bring in a document. If you want to use a font that's not resident in the imagesetter, you'll probably have to bring it along so your service bureau can download it.

You may notice certain typefaces on your Macintosh font menu that are named after cities, such as Chicago, Geneva, and Monaco. These fonts, which are installed when you first purchase your Macintosh, were developed for Apple's ImageWriter dot-matrix printer. They were not intended for high-resolution output and should be avoided in documents intended for a PostScript laser printer or imagesetter.

## Font Installation

Non-resident printer fonts are installed in the system in the same manner as any program. You simply drag the font files from the source diskette into your system folder. When you print a document containing the non-resident font, the Macintosh software downloads it to the printer or imagesetter. Some programs give you the option of removing the downloaded font after the print job or keeping it active until you shut down the output device.

## Screen Fonts

Along with your printer fonts, you need to install screen fonts so you can see the type displayed on your monitor. These are installed in the form of files known as "suitcases." When you open a folder containing a screen font, you actually see the icon of a suitcase with the letter "A" on it. To load the font into Macintosh systems prior to version 7, you use a program supplied by Apple known as the Font/DA Mover (Figure 4-12). This program copies fonts from the suitcase into the system, and also lets you copy fonts from one suitcase to another. You can give a suitcase file

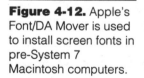

**Figure 4-12.** Apple's Font/DA Mover is used to install screen fonts in pre-System 7 Macintosh computers.

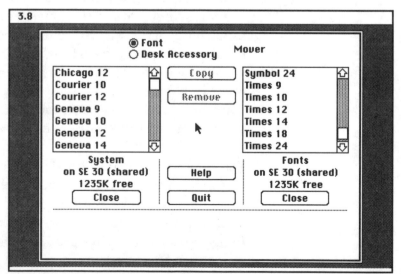

**WORKING WITH FONTS**

any name you want and add or delete fonts as you need them. Your Macintosh utilities manual tells you exactly how to do this. If you have System 7 or a font-management utility such as Suitcase II, installing fonts is as easy as dragging the icon to the proper folder.

If you purchase a PostScript clone printer, you may find that you do not need to install new screen fonts. Most PostScript clones are sold with renamed versions of Adobe/Linotype fonts that have been created to match the designs and metrics of the originals. Bitstream, for example, offers Dutch, Swiss, and Zapf Calligraphic in place of Linotype's Times, Helvetica, and Palatino. If you already have screen fonts for Times, Helvetica, and Palatino installed in your system, they will probably work with the clone fonts sold with the printer. When you specify "Times" in your publishing software, the printer substitutes the clone version of the typeface.

On the other hand, there are screen fonts and then there are screen fonts. Apple and Adobe, for example, both offer screen font versions of the same printer fonts that have minor spacing variations from one another. Your Macintosh system probably came with the Apple versions, but it is likely that your service bureau has the Adobe versions, in which case you may need to install the latter. You should check with the service bureau to make sure. They may be able to provide you with free or low-priced copies of the Adobe screen fonts.

Until recently, all screen fonts, even those for PostScript devices, were bit maps. Bit-mapped fonts, as we have seen, must be installed in the sizes and styles you most commonly use. If you specify a font of a different size, the Macintosh display software enlarges the closest screen font as best it can. The problem with this approach is that bit-mapped screen fonts, unlike scalable outline fonts, tend to take on a jagged appearance when enlarged or reduced.

The font looks fine when the page is printed, but it may be hard to read on the screen.

All this has changed with the advent of outline fonts and software packages like Adobe Type Manager. When you specify an odd point size on a system with ATM installed, the software creates a scaled rendition of the font without the jagged appearance. It does this in much the same way that PostScript creates type on a printer. You must still install at least one bit-mapped version of the font along with the Adobe software, but you don't need to load multiple point sizes.

Several software products are available that can assist in managing your fonts. One of the most useful is Suitcase II (Figure 4-13), a package from Fifth Generation Systems that makes it easy to install and remove fonts and desk accessories (DAs). The product itself is a DA, a special kind of utility program that can be launched from within other Macintosh applications. Desk accessories are found in the Apple menu along with system utilities like the Macintosh Chooser (which is used to select your output device). When you run a DA from the Apple menu, it pops up on the screen in the middle of your application. When you're finished

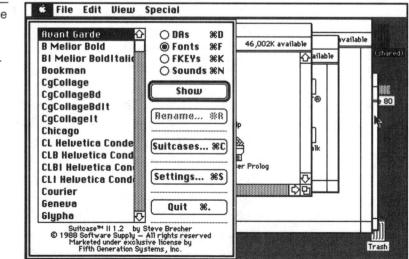

**Figure 4-13.** Suitcase II is a software utility that makes it easy to install Macintosh fonts.

with the DA, the system returns you to the original application.

Suitcase II, along with a similar program from Alsoft called Master Juggler, allows you to link suitcases in different folders and disk drives to the system folder. Under normal conditions, you can only select fonts that have been installed in the system folder. With Suitcase II and similar programs, you can use fonts installed in any folder — even on a separate disk or file server. When using the application on a LAN, you can install suitcases containing screen fonts for particular users on the file server. These users can select the fonts as if they've been installed in the system folder. The program also provides access to any printer font placed in the same folder as a suitcase containing the corresponding screen font.

*Font ID Conflicts*

One problem that sometimes arises with Macintosh fonts is the "Font ID conflict." A font ID conflict can occur whenever you print a document from one Macintosh that was created on another. For example, you create a 30-page document on your Macintosh using the Baskerville typeface for body text. But when you print it on your neighbor's laser printer, it comes out in Souvenir or some other undesired font. The problem can be particularly devastating when you try to print a document on a service bureau's Linotronic imagesetter.

Font ID conflicts occur because of the way the Macintosh system identifies fonts. An explanation of how this happens gets a bit technical, but understanding the problem can help you correct it if it arises.

Suitcases containing screen fonts also hold other data needed to identify and display the typefaces. This data is contained in a special kind of font file called a FOND, which stands for Font Family Descriptor. True to its name, the FOND contains information about specific families of typefaces, such as Times or Helvetica. This information includes width and kerning tables, along with a directory showing the sizes and styles of screen fonts in the family along with an identification code for each.

When you specify a typeface in a Macintosh program, the system looks for the FOND that corresponds to its family. The FOND provides the identification number for the file containing the selected screen font. If you select Souvenir, for example, the program sees it not as Souvenir, but as an ID number. When the document is printed, the printer replaces the selected screen font with the corresponding printer font.

The ID problem arises because of the numbering schemes used in the Macintosh system. In the original Macintosh system, room was available for just 256 font ID numbers. Since Apple reserved the first 128 numbers for its own use, this left 128 IDs for everyone else. This meant that you could have a maximum of 128 fonts installed on your system beyond those resident in the LaserWriter. With thousands of fonts available from dozens of vendors, it was inevitable that some would have duplicate ID numbers. If you copied a font into your system and already had a font with the same ID number, Font/DA Mover would renumber it. This means that there is no correspondence between fonts on one user's system and those on another.

*NFNTs*
Apple addressed this problem by developing a new kind of structure for identifying fonts. Fonts created under the new structure are called NFNTs, and can be identified by up to 32,767 numbers. It also encouraged software developers to identify fonts by name rather than number. But trouble remains. Some software developers have refused to identify fonts by name, meaning that font ID conflicts can arise if you use their packages. Other applications cannot recognize font IDs above number 511. It is likely that these problems will eventually be ironed out, but in the mean-

time, you should to check with your service bureau to identify any potential ID conflicts.

Several font utility programs can assist in working out font ID conflicts. Some allow you to convert older FONTS (in this case, the term is used to describe fonts numbered under the old scheme) to NFNTs. These include Fontographer, Suitcase II and Master Juggler. Varityper offers two free DAs, FontMaster and Font Wizard, that allow you to see ID numbers for the FONTs and NFNTs in the system.

## IBM-Compatible Computers

The IBM environment could be characterized by the exclamation "Every man for himself!" PostScript's hardware independence aside, each software package for IBM compatibles has its own method for installing and using screen and printer fonts. Not that the situation is total chaos. One function of graphical operating environments like Microsoft Windows and GEM is to handle generation of screen and printer fonts in a consistent manner. Once you install fonts for Windows, for example, they should work with any Windows application.

Font vendors have also provided some consistency by making sure that their products work with a wide range of applications. For example, Bitstream's Fontware, a program for installing and managing screen and printer fonts, is available in versions for GEM and Windows. Any version of Fontware will work with any Bitstream typeface.

Unlike Macintosh fonts, which are displayed as graphical icons, fonts on IBM-compatible computers are distinguished by their sometimes rather cryptic names. A Bitstream PostScript font, for example, may look like this:

cquq.pfa

"CQ" is an abbreviation of the typeface name, in this case,

"Bitstream Charter Roman." "U" refers to the character set, in this case, the standard U.S. character set. "Q" refers to a code for the output device, in this case a PostScript laser printer. "PFA" is Bitstream's standard file extension for PostScript outline fonts.

As we've seen with the Macintosh, identifying font files by name has some advantages. If you set a block of type in Caslon bold and send a PostScript file to a service bureau with that typeface, you can be pretty sure you won't see your document produced in Century Old Style.

As with the Macintosh, you may need to install screen fonts to ensure proper display of your printer fonts. But depending on the software you use, it may not be necessary. Ventura Publisher, for example, has two "generic" screen fonts, one serif, one sans serif. When you select a serif typeface, Ventura uses the serif screen font. Often, however, you get better on-screen type quality when you use the screen font supplied by the font vendor.

Whether you use an IBM or Macintosh computer, effective communication with your Linotronic service bureau will ensure that you manage your fonts effectively. Obviously, you need to know what fonts are installed, but beyond this, you need to know what resources are needed by your service bureau to get the most trouble-free output. If you're a Macintosh user, you may have to copy your fonts to a disk along with your document. Or you may have to install special screen fonts provided by the service bureau. Speak with your service bureau manager to find out.

# CHAPTER 5

# Working With PostScript Files

Look at the Mona Lisa or some other famous painting and imagine trying to describe it to a friend over the telephone. You could say that it's a picture of a dark-haired woman with a mysterious smile, but that wouldn't be much help if your friend was trying to recreate the masterpiece. No, you would have to give step-by-step instructions, from the top left corner to the bottom right, that would allow your friend to draw an exact replica. They say that a picture is worth a thousand words, but when you think about it, a thousand might be a pretty conservative estimate.

When you print a page on the Linotronic imagesetter, the PostScript page description language is performing a feat not unlike the task above. After you give the "Print" command from your desktop publishing software, your document is converted into a series of PostScript commands and sent to the imagesetter's RIP. The RIP follows these instructions to build an image of the page, which is then reproduced by the printer or imagesetter.

Normally you don't see the PostScript instructions as your page is being printed. The process is transparent, from Print command to final output. But you can also produce a PostScript program in the form of a computer file. In most cases, a PostScript file is a simple ASCII text file that contains PostScript commands. You can open it with a word processor to reveal the instructions that created your

**Figure 5-1.** When a
software package
prints to a PostScript
output device, it
actually creates a
PostScript program.

```
%!PS-Adobe-2.0
%%Title: ch4-2
%%Creator:  PageMaker 4.0  rocky
%%CreationDate:  9-7-1990, 12:30:45
%%For: Steve
%%BoundingBox:  0 0 612 792
%%Pages: 1 0
%%DocumentPrinterRequired: "" ""
%%DocumentFonts:  (atend)
%%DocumentSuppliedFonts:  (atend)
%%DocumentNeededFonts:  (atend)
%%DocumentNeededProcSets: AldusDict2  209   55
%%DocumentSuppliedProcSets:
%%DocumentPaperSizes: Letter
%%EndComments
%%BeginFile:  PatchFile
userdict /AldusDict known {(A previous version PageMaker
header is loaded.) = flush} if
%%EndFile
%%IncludeProcSet: AldusDict2  209   55
%%EndProlog
AldusDict2  begin
%%BeginSetup
letter
 mark
{
mark
                              .
                              .
                              .
statusdict /waittimeout 300 put
2550 3300 true false false BEGJOB
300 SETRES
25000 S_WORKING
save /SUsv exch def
%%EndSetup
%%Page: 1 1
BEGPAGE
SURSTR
false S_LOADFONT
(Times-Bold) FTRECODE
SUSAVE
                              .
                              .
                              .
/|_____Helvetica 140 100 mul 1000 div 980 16  0.00 0.00 0.00
1.00 (Black) false SET
2013 994 0.0000 0.0096
(Counter) 7 0 207 OUT
ENDPAGE
ENDJOB
end
%%Trailer
%%DocumentFonts: Times-Bold
%%+ Helvetica
%%DocumentSuppliedFonts:
%%DocumentNeededFonts: Times-Bold
%%+ Helvetica
%%EOF
```

publication (Figure 5-1). If you understand PostScript programming, you can even add your own instructions, or write a PostScript file from scratch.

# Benefits of PostScript Files

PostScript files serve two primary functions. First, they provide a means of transport for publications created by a desktop publishing system. Instead of printing a document on a laser printer, you can create a PostScript file, copy it to a diskette, take it to your service bureau, and have it produced on a Linotronic imagesetter. Because of PostScript's device independence, you can be sure that your pages will be printed correctly no matter what program was used to create them—if you take the proper precautions.

The second function is made possible by a special variety of PostScript known as the Encapsulated PostScript format (EPS or EPSF). EPSF files are used to store images created by sophisticated graphics programs like Adobe Illustrator, Aldus Freehand, or Corel Draw. An image created with one of these packages can be converted to EPSF format and used with other software. Because they store images in the form of PostScript commands, EPSF files are resolution independent. Print an EPSF file on a laser printer and you get 300-dpi output. But print it on a Linotronic imagesetter and you can get 1270-dpi output or more.

Another advantage is that EPSF files can contain almost any kind of image from a simple line drawing to a full-color photograph. PostScript is an object-oriented graphics format, which means that it builds images as a collection of objects that are described in mathematical terms. But PostScript files can also contain bit-mapped images, which are composed as an array of dots. These can be single-bit images — that is, black and white — or multiple-bit images in which each dot can be any one of many colors or shades of gray.

PostScript files are useful, but like everything else related to PostScript, they can often be a challenge to work with. For one thing, PostScript files tend to be big. Even a single magazine page stored in the form of a PostScript file can consume a megabyte of disk space or more. And even given PostScript's device independence, you must often make allowances for the kind of printer or imagesetter that will be used to produce a PostScript file.

In this chapter, we'll discuss these two kinds of PostScript files. First we'll cover the issues involved in producing PostScript files for output. Then we'll discuss the use of EPSF files for exchanging images between one program and another.

## PostScript Files for Output

If you plan to produce your documents at a service bureau, you will probably get to know the PostScript file format very well. Unless you use the service bureau's publishing software to produce your publications, you will probably create your document on your own computer, print it to disk as a PostScript file, and take it to the service bureau. There it will be downloaded to the imagesetter and produced at high resolution.

When you create a PostScript file from a desktop publishing or graphics package, the software is actually writing a PostScript program. All text and images on the page are converted into PostScript commands by means of a PostScript driver. Usually this driver interacts directly with the printer, but it can also produce PostScript output in the form of a disk file.

## File Sizes

One problem with these files is that they tend to be big. PostScript is extremely economical in the commands it uses to describe an image, but it still takes a lot of English-like instructions to draw a picture. This holds true for pages that consist mostly of text, but the biggest PostScript files are those that contain graphic images, especially bit-mapped images in color or multiple shades of gray.

To understand why this is so, remember that a bit-mapped image consists of tiny dots. If it is a scanned image, it might have as many as 300 dpi, which translates into 90,000 dots in a square inch. An image measuring 2 by 3 inches would thus contain 540,000 dots. Now suppose you have a gray-scale image in which each dot can be any one of 256 shades of gray. It takes eight bits of information — one byte — to describe each dot. This increases the total file size eight-fold. And that's just for the image portion of the page. Text and other graphic elements would add even more to the file size.

This is something of a worst-case scenario, because you would rarely want to scan a gray-scale image at 300 dpi. But even at 100 or 150 dpi, a gray-scale image in a PostScript file can quickly eat up your disk.

A standard PostScript file can contain several pages. But because of this tendency to eat up disk space, you may want to produce your PostScript files one page at a time. Another way to reduce disk space is to use a file compression program such as Stuff-It, which reduces the size of PostScript files by as much as 60 percent.

## File Transfer

PostScript files can be transported by several means. Service bureau users generally transmit them over phone lines by means of a modem. They can also be transported via diskette. These can be the 1.2 megabyte diskettes used in AT and 386 computers, or the 800K diskettes used on the

Macintosh. Iomega's Bernoulli Box uses removable 20- and 40-megabyte hard disks that can be used to transport PostScript files to a service bureau. The problem here is that the service bureau and client must both have compatible disk drive units.

## File Structure

Unless your PostScript file is in EPSF format, you can open it with a word processor or text editor and see the commands used to produce the page. These commands conform to a strict structure specified by Adobe Systems in its programming guides. If the structure is altered, or if any mistakes are made in the programming code, the page will be printed incorrectly or not at all.

### Header

All PostScript files begin with a "header" that provides essential information about its origin and contents (Figure 5-2). Some of this information must be present for the file to be printed. Other commands are optional. Header information can include a title, the date and time the file was created, the version of PostScript used to create it, and the author of the program. If the PostScript file was created by a software package like Adobe Illustrator, the program is listed as the author. Another important piece of information is the bounding box. This is a set of numbers that

**Figure 5-2.** The "header" is an important part of a PostScript file.

```
%!PS-Adobe-2.0
%%Title: ch4-2
%%Creator: PageMaker 4.0 rocky
%%CreationDate: 9-7-1990, 12:30:45
%%For: Steve
%%BoundingBox: 0 0 612 792
%%Pages: 1 0
%%DocumentPrinterRequired: "" ""
%%DocumentFonts: (atend)
%%DocumentSuppliedFonts: (atend)
%%DocumentNeededFonts: (atend)
%%DocumentNeededProcSets: AldusDict2 209  55
%%DocumentSuppliedProcSets:
%%DocumentPaperSizes: Letter
%%EndComments
```

indicate the overall dimensions of the page. Header comments are always preceded by two percentage signs: %%. The %%EndComments command indicates the end of the header information.

*Imaging Commands*

After the header come the commands that produce the actual image. This book makes no pretense to being a PostScript programming manual; several good ones are available. But as you can see, it consists of commands and numeric coordinates that describe how an image is to be constructed. Some of the commands may seem cryptic, but in many cases they are spelling out a relatively simple operation. For example, the commands to draw a square would amount to saying: "Go north 12 paces, then west 12 paces, then south 12 paces, then east 12 paces." Another command might instruct the program to fill the square with a 10-percent gray shade.

*Fonts*

Any PostScript file that contains text must include information about the fonts used in the document. Fonts can be included as part of the PostScript file. If not, the file must include a command that downloads any fonts that are not resident in the output device. In many cases, a program that produces a PostScript file will include references to all fonts in the system, not just those used in that particular document.

One of the most important commands in any PostScript file is the *showpage* operator. After the instructions to construct the page are processed, the *showpage* command tells the RIP to go ahead and produce the page on the printer or imagesetter. If the *showpage* command has been deleted from a PostScript file, it will be unable to produce a page.

# Producing PostScript Files

Each program has its own method for producing PostScript files. This is true even on the Macintosh, which is often distinguished by its consistency in implementing common functions across different programs. It is especially true in the IBM-compatible environment, which has always been characterized by a certain degree of anarchy in the way various software packages handle their printing functions.

## Three Methods

Programs generally produce PostScript files in one of three ways: through the printing function, the file save function, or a file export function. In most cases, the latter two functions are used by graphics packages to create EPSF files for exchange with other programs. Adobe Illustrator, for example, saves files in EPSF as its native format. Corel Draw, a popular IBM drawing package, uses a file export function to create EPSF files (Figure 5-3). As we'll see below, these EPSF files can be used to print pages and exchange graphic images with other software packages. As an output format, EPSF files differ from standard PostScript files in just one respect: while PostScript files can contain a range of pages, EPSF files can produce just one page at a time.

**Figure 5-3.** Corel Draw produces PostScript files through the file export dialog box.

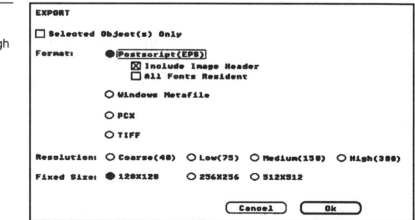

**WORKING WITH POSTSCRIPT FILES**

**Figure 5-4.**
PageMaker includes a
"Print to disk" function
that allows you to
create PostScript disk
files.

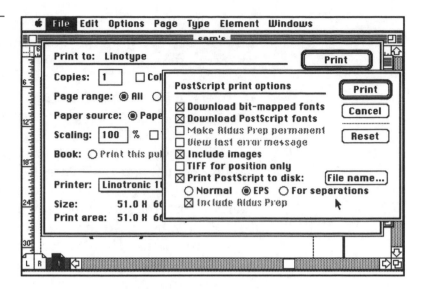

Most publishing programs, including Aldus PageMaker and Xerox Ventura Publisher, produce PostScript files as part of the printing function (Figure 5-4). You give the Print command, but instead of printing to an actual output device, you print to a PostScript file. Some graphics programs can also create PostScript files in this manner. Corel Draw, for example, allows you to print a PostScript file to disk in addition to creating EPSF files. Unlike the EPSF files created by Corel Draw, these "print to disk files" cannot be used to import an image into another program. Don't assume, however, that any graphics package that prints to disk can automatically create a PostScript file. GEM Artline, for example, produces PostScript files by means of an explicit PostScript command. The program's print to disk function produces a disk file that is not compatible with PostScript devices.

In most packages, "Print to disk" is an option when you give the Print command. For example, when you open the Print dialog box in Aldus PageMaker, an Options button opens a second dialog box that allows you to specify that the document be produced as a PostScript disk file. When you choose this option, you are prompted to enter a file name, after which the program generates the appropriate PostScript file.

Ventura Publisher, considered by many to be the leading publishing package on IBM-compatible computers, uses a somewhat different approach. Here, the "Print to Disk" option is found in the Set Printer Info dialog box under the Options menu (Figure 5-5). After selecting "PostScript" as your output device, you can choose from a number of interface ports that the printer can be connected to: LPT1: and LPT2: for parallel connections, or COM1: or COM2: for serial connections. The last "port" listed is "Filename." When you select this, the program is set up to print any documents in the form of a PostScript disk file. When you give the Print command, the program prompts you for the name of the file you want to create and produces it. Its default file extension for PostScript files is "C00," but you can change this if you want.

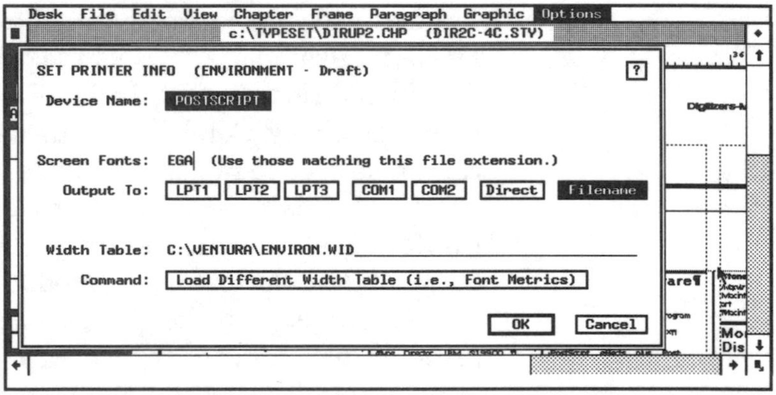

**Figure 5-5.** Ventura Publisher offers the option of printing to disk through its "Set Printer Info" dialog box.

## Font Management

One important consideration when producing a PostScript file for output is font management. If you try to print a document containing a certain typeface on a PostScript device that does not include that font, the printing software will substitute Courier. You must thus be sure that any fonts called for in your document are present in the output

**WORKING WITH POSTSCRIPT FILES**

device. If not, the fonts must be included in the PostScript file or downloaded by means of a command in the PostScript file.

Most publishing packages handle this font downloading automatically. But in some cases, you may have to specify how fonts are to be handled. In Corel Draw, for example, users are provided with a set of fonts named after famous cities that are equivalent to the Adobe/Linotype fonts used in most PostScript output devices. Corel's Avalon font, for example, is equivalent to Adobe's Avant Garde. If you want to print a Corel Draw document on a Linotronic imagesetter with Adobe/Linotype fonts, you need to check the "all fonts resident" option after you give the print command. When you do this, Corel Draw replaces its own typefaces with the equivalent Adobe fonts as it creates the PostScript file. But if the Adobe fonts are not present in the imagesetter, it will replace them with Courier. The alternative is to keep the Corel fonts in the document and manually download them to the imagesetter.

In addition to downloading fonts, a PostScript file created by a publishing or graphics package can include most of the same effects found on a printed page. Some packages allow you to produce multiple versions of a document as color separations or overlays. As we'll see in Chapter Seven, a color image generally consists of four components in varying intensities: cyan, yellow, magenta, and black. If your publishing or graphics package supports color separations, it will produce one PostScript file for each of the four color components. Some programs also allow you to enlarge or reduce the page as it is printed in the PostScript file. Corel Draw provides the ability to produce a reversed — negative—version of the page. This can be useful if you want to produce film negatives on a Linotronic imagesetter.

### Crop Marks
One helpful feature is the ability to produce PostScript files that include crop marks or crosshairs (Figure 5-6). These

**Figure 5-6.** Some programs allow you to add crop marks when you produce an EPSF file.

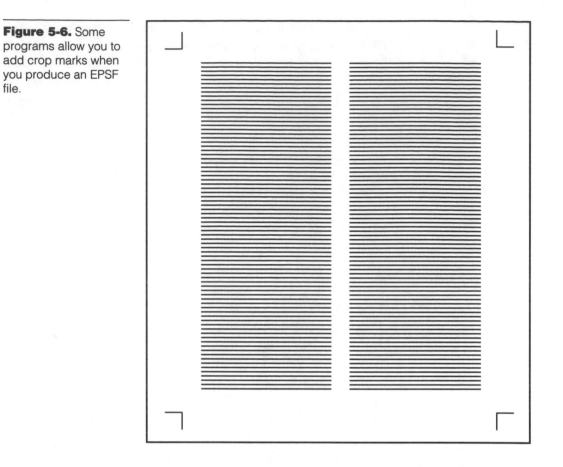

allow your printing press operator to align the page correctly. In Ventura Publisher, for example, you use the Print dialog box to specify that crop marks be included in the document. If the program has been set up to produce a PostScript disk file, it will add the crop marks as part of the image of the page.

# Downloading PostScript Files

Once a PostScript file has been created, it can be produced on a laser printer or imagesetter by a process known as downloading. The file is copied to the PostScript RIP, which processes the PostScript commands as if you were printing directly from a publishing or graphics program.

Several methods are available for downloading PostScript files. If you have an IBM-compatible computer that is directly connected to the imagesetter, you can issue a simple DOS copy command. For example, if the imagesetter is connected to the first parallel port (commonly known as LPT1), the following command will download the file to the output device:

```
COPY FILE.PS LPT1:
```

If your computer is part of a LAN, you can use the network software to download the file. The TOPS network, for example, includes a printing utility called NETPRINT that allows you to download a PostScript file to any printer or imagesetter on the network. Once the software has been correctly configured, you simply type PRINT followed by the file name.

Macintosh users can download PostScript files by means of a downloading program (Figure 5-7). These programs are generally quite easy to use. First, you use the Macintosh system Chooser to select the printer or imagesetter on which you want to produce the file. Then you simply choose the downloading option and double-click on the name of the file you want to print.

# EPSF Files

The Encapsulated PostScript format is a special kind of PostScript file that allows images to be exchanged among various software packages. Like any PostScript file, an EPSF file uses PostScript commands to build an image. But it also has a number of special features that distinguish it from standard PostScript files used to create a page.

### Origin of EPSF

Many people are surprised to learn that the EPSF format was not created by Adobe Systems, which developed the

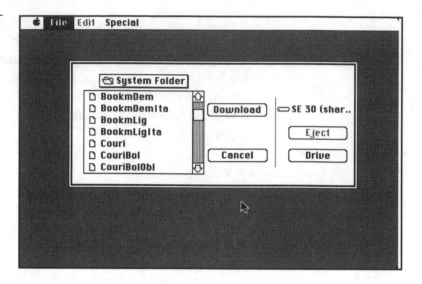

**Figure 5-7.** The Macintosh Font Downloader allows you to copy PostScript files or fonts directly to the printer.

PostScript language. Instead, it was created by Altsys Corp., a Dallas-based company that developed the Freehand illustration program that is sold by Aldus Corp. Altsys was seeking a file format that would allow different graphics packages to exchange documents, and found that PostScript offered many advantages as a graphics file format. Its resolution independence meant that EPSF files could be printed at the maximum resolution of the output device. Its versatility meant that EPSF files could contain any kind of graphic image, even a color photograph.

The EPSF format quickly caught on with software developers and end users, and it has become one of the most popular graphics formats for high-resolution images. Almost all professional-level illustration programs, including Adobe Systems' Illustrator, Aldus Freehand, and Corel Draw, can create EPSF for export to other programs. Most can also import EPSF files to use as part of other images. Almost all leading publishing programs, including Aldus PageMaker, Xerox Ventura Publisher, Design Studio, and QuarkXPress, can import files in EPSF format. A new generation of color publishing products uses the EPSF format to produce color separations.

The key difference between an EPSF file and a standard PostScript file is the use of a bit-mapped image for display purposes. Converting PostScript commands into an actual image takes a lot of processing power. Most software packages just don't have the display capabilities needed to show a PostScript image on the screen. But when you create an EPSF file, most programs will also produce a low-resolution bit-mapped version of the image that is embedded in the file.

On the Macintosh, this image is produced in the form of a PICT file, a popular graphics format created by Apple Computer. On IBM compatible computers, the bit-map is provided in the form of a TIFF file, another popular format developed by Aldus Corp.

### Importing EPSF

An EPSF file that includes an embedded bit-map can be imported into a publishing or graphics package like any other image file. You can even import an EPSF file into a program used to create another EPSF file.

An image in EPSF format can be moved, cropped, stretched, rotated, enlarged, or reduced on screen, but you cannot change its actual contents. When you print your document, the software discards the bit-mapped image seen on the screen and substitutes the actual high-resolution PostScript image, scaled or stretched as the user desires.

Even if an embedded bit-map is not included in an EPSF file, you can still use the image. An EPSF file without a bit-map appears as a gray box that corresponds to the image area (Figure 5-8). The box may also include written comments from the PostScript header that indicate how it was created, but you cannot see a representation of the image. Even so, the EPSF file will be correctly printed when you produce the page.

For a program to display an EPSF file without the gray box,

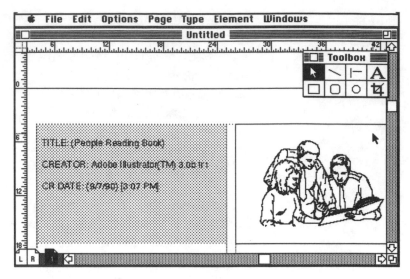

**Figure 5-8.** If an EPS file does not include a PICT or TIFF screen representation, it will be displayed as a gray box.

it must be able to import the file format used to create the bit-map. Almost all Macintosh packages can import PICT, and most IBM-compatible packages support TIFF.

EPSF files differ in other ways from standard PostScript files. These differences allow PostScript to do something for which it was not originally intended: to be inserted into another file rather than printed by itself. For this reason, EPSF files include PostScript comments that allow their use within documents created by other software packages.

Some PostScript commands act on the entire page rather than a particular image and thus cannot be used in an EPSF file. For example, you can imagine what the *"erasepage"* operator would do if it were included in an EPSF file. The *"quit"* operator can have similar consequences.

One page-oriented command that can be found in EPSF files is *"showpage."* Any software package capable of importing EPSF files will remove this command when the file is imported.

EPSF offers numerous advantages as a graphics file format. It works with a wide range of software packages that produce high-quality, high-resolution artwork. It is one of the few formats that can describe almost any kind of image, from a simple black-and-white line drawing to a full-color photograph. Once it's imported into a publishing or graphics package, it can be stretched, enlarged or reduced without loss of image quality.

### Disadvantages of EPSF

But EPSF files also have their disadvantages. As with standard PostScript files, EPSF files can consume large amounts of disk space, especially if they contain gray-scale images. For this reason, you should use the EPSF format only when there is not a suitable alternative. In general, EPSF files are used for object-oriented graphics produced by illustration packages like Illustrator, Freehand, or Corel Draw. You can store a scanned image in EPSF format, but you are generally better off using TIFF.

Another disadvantage is that images in EPSF format are difficult to modify. Most publishing and graphics packages can stretch, rotate, or re-size an EPSF file, but they cannot actually change its contents. EPSF files in a straight-text format can be edited in a word processor, but you must have a strong command of PostScript programming to make any but the most modest changes. Some packages, such as PS Tutor and LaserTalk, help simplify the task of PostScript programming by showing you the effects of PostScript commands as you enter them. Still, PostScript programming is something that most users will want to avoid.

### Creating EPSF Files

Each program has its own method for creating EPSF files. Some illustration packages, such as Corel Draw or GEM Artline, create EPSF files through an Export function. Others, such as Illustrator, create EPSF files through the file save function.

Some packages offer more control over the structure of

EPSF files than others. Illustrator, for example, provides several EPSF options: you can create files with a TIFF bit map for IBM compatibles, a PICT bit map for the Macintosh, or with no bit map at all. Corel Draw allows you to determine the resolution of the embedded bit map. The default resolution is fixed at 128 by 128 dots, but this can be increased to 256 by 256 or 512 by 512. Alternately, you can choose to have the bit map displayed at 40 to 300 dpi.

Many illustration packages have special features that take advantage of PostScript's graphics capabilities. For example, Corel Draw and Micrografx Designer allow you to fill objects with a number of PostScript shading effects. These include bricks, bars, birds, bubbles, craters, leaves, stars, spider webs. Each effect is an actual PostScript program that can be modified by entering values in a dialog box. You cannot see the effects on the computer display, but they will be printed as part of the final image.

Some publishing packages can produce EPSF files in addition to standard PostScript files. This is helpful if you want to use an image of a page as a piece of artwork. In PageMaker, for example, you can create EPSF files with embedded bit maps in TIFF or PICT formats. Pages produced in PageMaker can thus be treated as graphic images. Ventura Publisher creates PostScript files that can be imported into a Ventura file, but they are not true EPSF files and cannot be used with other programs.

An innovative Macintosh program called SmartArt allows users to perform many interesting effects with PostScript files (Figure 5-9). The program is sold with a series of PostScript programs that produce certain text and graphic effects, such as a line of text that is slanted and banded. Using the program's commands, you can manipulate the effect in many ways, such as by changing the fonts, text, style, rotation angle, and foreground or background shading. The program shows the effects of your changes by sending the file to a PostScript printer, then displaying the

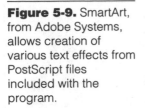

**Figure 5-9.** SmartArt, from Adobe Systems, allows creation of various text effects from PostScript files included with the program.

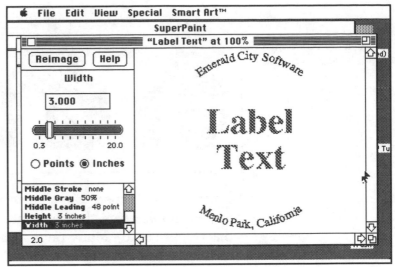

results on screen. Once you are satisfied with the image, you can save it in EPSF or PICT formats. The package only works with the PostScript effects provided by the developer.

## Using EPSF Files

Many publishing and graphics packages can import EPSF files. In general, these programs handle EPSF files in the same way they handle other object-oriented graphic images. You can perform certain operations, such as scaling and cropping, but you cannot change the actual content of the image. This is true even if you import an EPSF file into the program that created it in the first place.

*Scaling*

"Scaling" means to enlarge or reduce an image. Unlike bit-mapped graphics, which are locked into a certain pattern of dots, graphics in EPSF format can be enlarged or reduced without degrading the image. For example, when you enlarge an EPSF image, any details lost at the smaller size will be revealed at the larger size.

To stretch or compress an image, you simply enlarge or reduce it in a single direction—horizontal or vertical— without altering the other dimension. In doing so, you will

be altering its "aspect ratio," a term that describes the relationship between the horizontal and vertical dimensions (Figure 5-10). Many publishing and graphics packages have "constraining features" that force the image to maintain its aspect ratio when it is scaled.

Most programs also allow you to crop an EPSF image. To crop an image is to trim away a portion of it. You might do this to change it from a vertical to a horizontal orientation, or to draw attention to a particular object in the image. Some graphics programs also allow you to rotate EPSF images, but this is not true with most publishing programs. In some cases, image rotation is limited to 90-degree increments. Like any graphic, an EPSF image can also be moved around the page layout by dragging it with a mouse or other pointing device.

### Other Modifications

We noted above that most graphics programs cannot modify the contents of an EPSF file. One of the circumstances under which you can modify an EPSF file is by doing your own custom programming. This is a difficult task, but it is made easier by a number of software packages designed as PostScript programming aids.

The first step in modifying an EPSF file is to convert it back into pure text format. Several utility packages are available for this purpose. Then you can open the file and make any necessary changes to the program instructions.

PostScript programming aids, such as PS Tutor and LaserTalk (Figure 5-11), provide a means of instant feedback for PostScript commands. These packages provide an on-screen display of a PostScript image. PS Tutor does this by means of a PostScript clone, while LaserTalk uses an actual PostScript printer to provide a representation of the image. As you enter or change PostScript commands, the software displays the change. The programs also include on-line references to the PostScript commands.

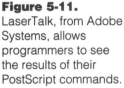

**Figure 5-11.**
LaserTalk, from Adobe Systems, allows programmers to see the results of their PostScript commands.

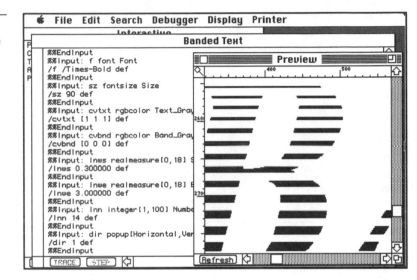

These are useful products, but PostScript programming is still best left to the experts. If you don't know what you are doing, you could easily introduce a "bug" that prevents the image from being properly printed or displayed. Fortunately, most illustration packages are powerful enough that you should be able to produce EPSF files that need little modification.

# CHAPTER 6

# Halftone Imaging

For many desktop publishing users, halftones are the major justification for producing output on a high-resolution imagesetter. Linotronic imagesetters do a good job of producing text and line art, but many users find that low-cost laser printers are adequate for many documents. This is not the case with halftones. A laser printer can generate halftones, but the output quality usually leaves much to be desired. If you want to produce halftones that match the quality found in professional publications, the imagesetter is really your only choice.

In this chapter, we'll explore the issues involved in producing halftones on Linotronic imagesetters. We'll begin with a discussion of traditional halftones produced by photographic means, and explain the terminology used to describe them. Then we'll describe the process of producing a digital halftone, from input on a scanner or digitizer to final output on the imagesetter. To keep things simple, we will limit the discussion to monochrome—black and white—halftones. In a later chapter, we will build on this knowledge to describe how images are produced in color.

## Traditional Halftones

If you look at a photograph in a newspaper or magazine, what you are really seeing is an optical illusion. At first glance, it appears little different from the kind of black and white print you might get from a photo processing lab. But

**Figure 6-1.** A
traditional halftone uses
variable-sized dots to
give the illusion of gray
shades.

look closer, and you can see that the image consists of tiny
dots, a little larger in dark areas, smaller in light areas.
Together, the dots give the illusion of varying shades of
gray. In the lexicon of printers and publishers, this illusion
is known as a halftone (Figure 6-1).

Without halftones, it would be nearly impossible to reproduce photographs. Black and white film contains thousands of tiny silver particles, each of which can represent
almost any level of gray. Because the particles are so small,
neighboring gray areas appear to blend smoothly with one
another. But printing presses cannot vary the intensity of
the ink they lay on the page. Dots can be black or white, and
that's it. However, if the photograph is converted into a
halftone, it can be reproduced quite easily.

In the traditional printing process, a halftone is created by
shooting a photograph—also known as a continuous-tone
image—through a fine screen. It can be a glass screen
placed between the camera and photograph, or a film-
based screen. What results is an image consisting of tiny
dots. The dots can be square, round, or oval, but they are
always spaced at even intervals. Dots in light areas of the
image are very small, while dots in dark parts of the image
are slightly larger. The varying dot sizes blend together

and fool the unaided eye into perceiving continuous-gray shades. The halftone ends up looking much like the original photograph.

Desktop publishing systems have introduced a new kind of image known as the digital halftone. Typically, the process begins when the photograph is converted into a digital format by means of a desktop scanner. It ends on a piece of resin-coated paper or film produced by a Linotronic imagesetter.

## Line Screen

The quality of a halftone depends on several factors, the most important of which are resolution and gray scale. Resolution is determined by how closely the dots in the halftone screen are spaced. Desktop publishing users are accustomed to measuring resolution in dpi. But halftone resolution—also known as "screen frequency"—is measured in lines per inch, or lpi. Halftones with a high screen frequency have a high density of dots—and present a sharper image (Figure 6-2).

In most cases, halftones printed in newspapers are limited to 65 to 85 lpi because of the low quality of the newsprint on which they are reproduced. Magazine halftones generally offer 120 to 150 lpi, while fine art reproduction generally requires 200 lpi.

**Figure 6-2.** Halftone screen frequencies are measured in lines per inch. Newspaper halftones (left) generally use a 65 to 85 line screen. Magazine halftones (right) use a 100 to 150 line screen.

## Gray Scale

Gray scale, the other major factor affecting halftone quality, is a measure of the different levels of gray that an image can display. A continuous-tone image can show a nearly infinite range of gray values. Halftones simulate these continuous tones by varying the size of halftone dots. The number of possible dot sizes determines how many shades of gray can be shown in a particular image. In an image with 64 levels of gray, a dot can be any one of 64 sizes.

To look acceptable, a halftone needs somewhere between 16 and 256 levels of gray (plus one if you want to count white as a gray level). Images with an insufficient number of gray levels will appear to have distinct bands of gray rather than smoothly blended tones in transitional areas (Figure 6-3).

In addition to varying screen frequencies, halftones can have different screen angles and dot shapes. Screen angle refers to the angle at which the screen was placed when the halftone was created. If the screen is placed horizontally over the photograph, the screen angle is zero degrees. If it's placed vertically, the screen angle is 90 degrees. A 45-degree screen angle, in which the screen is turned diagonally, has been found to be the most pleasing to the eye since it doesn't conflict with vertical and horizontal lines in the image (Figure 6-4). The dot shape refers to the shape of the tiny holes in the halftone screen. Dots can be round, square, or elliptical. Some halftone screens use lines in-

**Figure 6-3.** Images require sufficient variation in gray shades to look realistic. The image on the left has just 16 gray shades. The image on the right has 256 gray shades.

**Figure 6-4.** Halftones can have varying dot angles. A 45-degree angle generally produces the best results.

| 0° | 30° |

stead of dots. Most halftones produced by desktop publishing systems have round dots.

# Digital Halftones

Halftones produced on a Linotronic imagesetter are similar to those produced by traditional means. If you look at a digital halftone under a magnifying glass, it appears to consist of tiny dots of varying sizes, larger in dark areas and smaller in light areas. Like a traditional halftone, a digital halftone can have dots of varying shapes and angles. But the process for producing a digital halftone is much different than the traditional halftoning process.

Digital halftones can be produced on laser printers or high-resolution imagesetters. Like a printing press, these output devices are limited by their inability to produce dots with varying intensity. But they have an additional limitation not shared with the printing press, for they cannot vary the size of the dots they print. A dot produced on a 300-dpi laser printer measures about one-three-hundredth of an inch. A dot produced on a Linotronic 300 is also limited to the imagesetter's resolution, whether it's 1270 dpi or 2540 dpi. As we have seen, a traditional halftone gives the illusion of varying shades of gray by varying in size. To give the effect of variable-sized dots, digital output devices use a technique called "dithering."

## Halftone Cells

So far, we have discussed halftone dots, which can vary in size, and digital dots, which cannot. Dithering introduces a third kind of dot that we'll call a halftone "cell." When a page layout program sends a dithered image to a laser printer, it groups two or more printer dots into a cluster— the halftone cell—that simulates one halftone dot (Figure 6-5). To avoid confusion in referring to these different kinds of dots, printer dots are sometimes known as "spots."

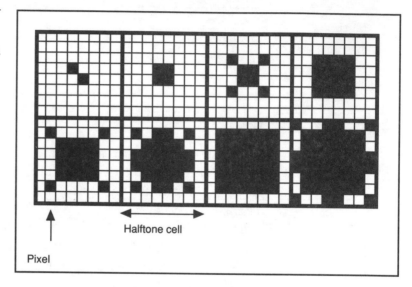

**Figure 6-5.** When a laser printer or imagesetter produces a halftone, it groups dots into halftone "cells" to simulate the variable-sized dots of a traditional halftone.

Halftone cell

Pixel

A halftone cell might consist of 64 spots arranged in an eight-by-eight square. The number of black printer dots determines the size of the halftone cell, and thus the amount of gray represented. If all printer dots are black, the halftone cell is black. If 48 out of 64 printer dots are black, the halftone cell appears dark (75 percent) gray. If 16 out of 64 printer dots are black, it's a lighter shade (25 percent) of gray.

With dithering we can get the appearance of gray scale, but at a cost in resolution. If you have a 300-dpi laser printer in which a four-by-four array is used in each halftone cell, resolution is cut—by a factor of four—to 75 dpi. This approaches the resolution you might find in newspaper photographs, but the 16 levels of gray provided by the clustering are below the minimum needed for realistic reproduction of an image. If you increase the number of gray levels to a more acceptable 64, effective resolution is reduced to about 50 dpi.

A laser printer with 300-dpi resolution produces barely acceptable halftones. But if you produce your pages on a Linotronic imagesetter, the resolution/gray scale trade-off becomes irrelevant. You could have halftone cells with 16 dots on a side, enough for 256 levels of gray, and still get a line screen of 150 lpi.

There is a relatively simple formula that can tell you the maximum screen frequency of a digital halftone given the number of gray shades and the resolution of the output device. Just take the square root of the number of gray shades and divide it into the resolution. If you have an image with 16 gray shades, the square root would be four. Divide four into 300, and you get 75—the maximum screen frequency for most laser printers. An image with 256 gray shades produced at 2540-dpi resolution on an imagesetter can have a maximum resolution of slightly less than 160 lines. The square root of 256 is 16. Divide 16 into 2540 and you get 158.75.

To keep things simple, we've been dealing with nice round numbers—and assuming that the halftone dots will be printed in straight horizontal lines. But as we noted above, publishers have found that halftones tend to look better when the dots are set off at 45-degree angles. This changes our numbers somewhat, but the trade-off between gray scale and resolution remains.

## Capturing the Image

The process of producing a digital halftone begins with a desktop scanner or video digitizer. A scanner is a hardware device that converts images on paper or slides into a digital format. A digitizer converts images from a still video camera, TV, or VCR into the same kind of digital format.

*Scanners*

In most cases, scanners are the preferred means of capturing images for use as digital halftones. But not any scanner will do. To produce a halftone, the scanner needs what is known as gray-scale capability. This means that it can recognize multiple shades of gray in the images it captures. Most gray-scale scanners sold these days can recognize up to 256 shades of gray. They are also known as "8-bit" scanners, since it takes eight bits of data to store up to 256 values. Older gray-scale scanners were limited to 4 bits (16 gray shades) or 6 bits (64 gray shades). Some scanners without gray-scale capability are advertised as being able to produce halftones. However, these should be avoided. They use a form of dithering on the input end that severely limits the quality of the images they produce.

Most gray-scale scanners have a flatbed design (Figure 6-6). They resemble small photocopiers, with a removable cover and glass platen. The user places a photograph on the platen, and an array of photosensitive sensors known as CCDs moves across the scanning area. Sheetfed or edge-

**Figure 6-6.** Flatbed scanner.

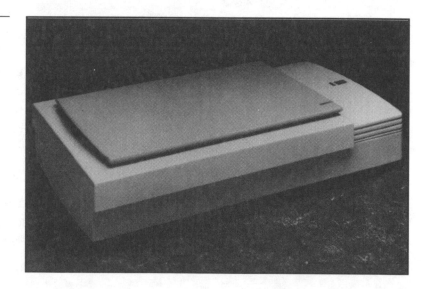

**Figure 6-7.** Sheetfed
scanner.

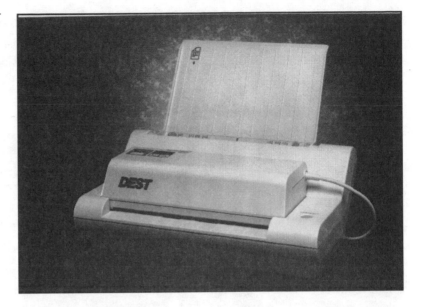

**Figure 6-8.** Hand-
held scanner.

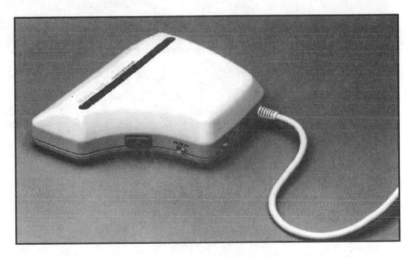

fed scanners use rollers to pull the photograph over the
CCD array, but these are not recommended for gray-scale
image capture (Figure 6-7). Some hand-held scanners
(Figure 6-8) have gray-scale capability, but again are
generally not suitable for quality image work. Slide scanners (Figure 6-9), which capture images on 35mm slides,
can be used to capture gray-scale images, but these are
quite expensive and are intended largely for color work.

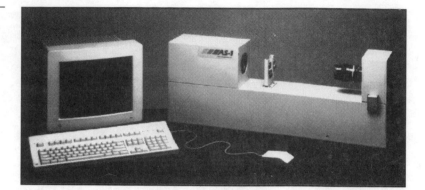

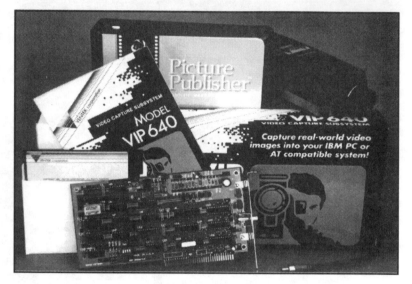

### Digitizers

Digitizers are generally used to capture real-life three-dimensional images (Figure 6-10). They typically have two components, a video camera and frame-grabber board installed in the computer (though the frame-grabber alone is often referred to as a digitizer). The video camera records the image, while the frame-grabber converts it into a digital format. Once the image is converted, it is identical to an image captured by a scanner. Video digitizers should not be confused with digitizing tablets, which allow users to draw images using an electronic stylus and tablet.

A digitizer offers the ability to produce instant halftone images, since you don't have to bother with photo processing. But low-cost digitizers are generally limited in the quality of the images they can produce. They are also unsuitable for capturing images on paper. Digitizers that offer high-resolution image capture are quite expensive. Some scanners use a camera-like mechanism to capture images, and thus appear to be scanner/digitizer hybrids. Again, these tend to be more expensive.

*Interfaces*

In the IBM environment, the scanner is connected to the computer by means of an interface board. In the Macintosh environment, most scanners are connected by means of the Small Computers Systems Interface—SCSI for short. Some scanners include a built-in SCSI interface, while others connect to the computer by means of a separate SCSI interface box.

*Scanner Control Software*

In addition to the interface, most scanners and digitizers are sold with software that allows the user to control the image-capture process (Figure 6-11). Sometimes this is a stand-alone program that handles scanning and nothing else. Other programs include scanner control as part of a comprehensive set of image-editing functions.

Some scanners include external controls for brightness and contrast, but for the most part, all scanning functions are controlled by software. When operating a scanner, users must concern themselves with settings for brightness, contrast, scanning area, scale factors, and resolution.

Brightness and contrast functions work much like the same controls on a TV set. Brightness adjusts the lightness or darkness of an image, while contrast adjusts the amount of difference between light and dark areas. The default settings for your scanner control software will usually work best, but sometimes you may want to make adjustments.

An image that appears too dark may benefit from raising the brightness level. You may be able to bring out greater detail in washed-out photos by increasing contrast.

The scanning area is the actual portion of the photograph you want to capture. As we shall see, gray-scale images consume a lot of disk space. One way to reduce this is to capture only that part of an image you really need. Typically, this is done by drawing a box over a representation of the scanning area displayed on the screen. To assist in this effort, most scanner control programs allow you to perform a low-resolution "test scan" to preview an image before using it. You can select the appropriate portion in the preview and scan it at full resolution.

Scaling factors allow you to enlarge or reduce a photograph as it is being scanned. Though you can also do this afterwards in your desktop publishing or image-editing software, you generally get better image quality if you handle scaling on the initial scan. In most cases, scaling is measured in terms of percentages. If you scan an image at 100 percent, it is reproduced at its original size. An image scanned at 200 percent doubles in dimensions. An image scanned at 50 percent is cut in half. Most scanner control programs allow you to reduce images as much as you want. But if you want to increase image size, you generally have to reduce the resolution at which the image is scanned.

## Determining Optimal File Size

When producing scanned images for output on a Linotronic imagesetter, you will be constantly faced with a trade-off between keeping the file to a reasonable size and achieving the best possible output quality. In general, the best strategy is to create a scanned image file that has as much image data as your output device can use. Since scanned images can take up a considerable amount of disk space and can slow the performance of an imagesetter, you don't want to create a scanned image with more data than can be appreciated by the human eye.

To determine the optimal file size for your image, you must know two things: the size of the final reproduced photograph, and the line screen of the reproduced photograph. The general formula for optimal file size is:

File Size = 2 x (Line Screen)$^2$ x (Size of Final Photo)

This formula assumes that you are using an 8-bit (256-gray level) scanner, which is highly recommended for producing images on a Linotronic imagesetter. With an 8-bit scanner, each pixel consumes 1 byte (8 bits) of data. Thus, a 100-lpi halftone consumes 100 x 100 = 10K bytes of data per square inch. (Note that there are 100 x 100 = 10,000 dots in a square inch; each dot consumes 1 byte). The factor 2 in the equation above results from the fact that you must provide more data than the bare minimum to compensate for various errors that come into the scanning process.

For example, imagine that you are going to scan a 4 x 5 inch photograph and reproduce it at the same size on a Linotronic imagesetter using a 100 line screen. The optimal size, using the formula above, is then:

$$2 \times (100)^2 \times (4 \times 5) = 400 \text{ Kbytes}$$

What's important to note is that you can arrive at the 400K figure by varying either the scanning resolution or scale factor, or both. In the example above, we could have arrived at a 400K scanned image by scanning at 200 dpi and 100 percent scale factor, or we could have scanned at 100 dpi and 200 percent scale factor, or 300 dpi and 67 percent scale factor — each produces the same result.

Many scanner control programs, including DeskScan from Hewlett-Packard, show you the size of the file you are about to create, which makes it easy to adjust scale factor and resolution to arrive at the required file size.

Suppose instead that we had started with an 8 x 10 photograph, but still wanted the final image size to be 4 x 5 (50 percent reduction factor). How big should the file be? It should still be 400K—we have not changed the final image size or the line screen. You will find, however, that you can achieve the 400K file size by scanning at a lower resolution. In general, if you are reducing the size of a scanned image, you can lower the scanning resolution; if you are enlarging the image, you must increase the scanning resolution.

Also, notice that the amount of disk space required for an image increases geometrically as the size of the reproduced image increases. A halftone enlarged to 200 percent size requires four times as much data as the original. It's also a good idea to crop your image as you scan, rather than within the desktop publishing program. Although programs like Aldus PageMaker and QuarkXPress allow you to crop scanned images after you have placed them on a page, you will still be carrying around excess (though unseen) image data within your publication files.

Table 6-1 gives the optimal file size for scanned images of various sizes and line screens.

**Table 6-1.** Optimal file size for various images.

| Line screen | Final Image size (sq. in.) | Required file size (KB) |
|:---:|:---:|:---:|
| 65 | 6 | 50 |
| 65 | 12 | 99 |
| 65 | 20 | 165 |
| 85 | 6 | 85 |
| 85 | 12 | 169 |
| 85 | 20 | 282 |
| 100 | 6 | 117 |
| 100 | 12 | 234 |
| 100 | 20 | 391 |
| 133 | 6 | 207 |
| 133 | 12 | 415 |
| 133 | 20 | 691 |
| 150 | 6 | 264 |
| 150 | 12 | 527 |
| 150 | 20 | 879 |

If you use a digitizer, the issue of resolution is pretty much irrelevant. Since the number of pixels a digitizer can render is fixed, all images will appear at the same resolution. Factors like lighting and focus that are common concerns in photography are much more important.

Our discussion so far has focused on scanning continuous-tone photographic images. It is generally not recommended to scan an image that has already been converted into a halftone. When the image is ultimately produced, the new halftone screen will conflict with the original screen, producing distortions known as moire patterns. If you absolutely must scan a halftone original, scan it as line art at full resolution in single-bit mode, and be careful not to scale it up or down.

### Saving the Image

After an image is scanned, it is saved as a graphics file in one of several standard formats. Some scanner programs first scan an image into the computer's memory, after which it is saved as a file. This can cause problems if you don't have enough memory to handle the image data. Other programs can scan the image directly into a disk file.

The format you choose to store your image largely depends on the software you want to use it with. Obviously, you should not save an image in a format that's not supported by your image-editing or desktop publishing software.

TIFF, short for Tagged Image Format File, was developed by a small group of hardware and software developers as a standard file format for saving gray-scale images. It is supported by most leading publishing programs on the Macintosh and can be used with PC-based software as well. TIFF files can store images with up to eight bits per pixel, allowing a full 256 levels of gray. It is usually the preferred format for saving gray-scale images, since it is well supported by most software packages.

RIFF is a proprietary gray-scale file format developed for Letraset, which markets DesignStudio, Ready,Set,Go, ImageStudio, and other popular programs. The major advantage of RIFF is that it allows compression of gray-scale images for reduced consumption of disk space. Its major disadvantage is that it works only with Letraset software. Users of Letraset packages may want to save their images in RIFF format, but others should choose TIFF.

PICT2 is an extension of Apple's PICT format for MacDraw and other draw-type programs. Unlike the original PICT format, PICT2 can save bit-mapped images with gray-scale information. However, even in the Macintosh environment, it is not as widely supported as TIFF.

Encapsulated PostScript, as we saw in Chapter Five, is a variant of the PostScript file format that can be used to store halftone images. However, it should be avoided unless you have no other alternative. EPSF files consume even more disk space than uncompressed TIFF files. And once saved in EPSF format, the image cannot be retouched or enhanced with image-editing software.

### Editing the Image

After an image is saved, you may want to import it into an image-editing program. These programs, which include ImageStudio (Figure 6-12) and Digital Darkroom (Figure 6-13) on the Macintosh and Picture Publisher and Gray FX on the PC, are designed specifically for manipulation of gray-scale images produced on a scanner. In many ways, they are similar to paint packages like MacPaint or PC Paintbrush, but are geared toward images in which each dot can be one of several shades of gray.

The better image-editing programs can produce photographic effects previously achieved only in a darkroom. By lowering the gray value of each pixel, you can make an image lighter. By reversing the gray values, you can

**Figure 6-12.**
Letraset's ImageStudio was one of the first gray-scale image-editing programs for the Macintosh.

**Figure 6-13.** Digital Darkroom, from Silicon Beach Software, is a popular image-editing program for the Macintosh.

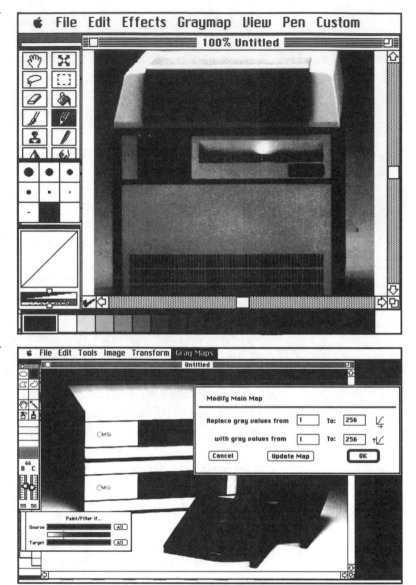

convert it into a negative. Painting tools allow you to remove objects or add new ones. Cloning and texture tools allow you to reproduce a portion of an image elsewhere. Selection tools allow you to copy an object from one image and paste it into another.

## Display Choices

To use these programs effectively, you need a monitor capable of gray-scale display. Many companies sell monitors for the Macintosh that can show up to 256 levels of gray. Most of these are large-screen monitors capable of showing a full page or two pages side-by-side. In the IBM environment, you need a VGA monitor for gray-scale display. Standard VGA monitors can show 16 levels of gray at 640 x 480 resolution, or 64 levels of gray at 320 x 200 resolution. Monitors with a Super VGA or 8514/A adapter can show 64 levels of gray at full resolution. Most PC-based image-editing programs can work with a full 256 levels of gray, even if the monitor is limited to 16 or 64 levels.

Another hardware requirement for many image-editing programs is extra memory, especially on the Macintosh. Some programs require a minimum of four megabytes, and preferably eight if you want to use them to full advantage. Other programs get around these substantial memory requirements by using what is known as virtual memory, in which portions of the image not being edited are stored on disk. However, even with these programs, you will find yourself working more efficiently if you can store the entire image in RAM.

## Editing Tools

Though image-editing programs vary widely in functionality, some features are common to all. These include painting tools, selection tools, image filters, and gray-scale editing functions. Gray-scale functions are especially useful in improving the ultimate quality of halftones produced on the imagesetter. They typically include brightness and contrast controls, histogram and equalization functions, and perhaps the most important of all, the gray-map—or gamma curve—editor.

## Gray Maps

The gray-map editor is a linear graph that shows how gray values in the original image correspond to values in the

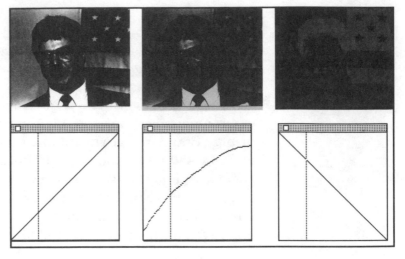

displayed image (Figure 6-14). In a normal image, the line runs diagonally from the upper right to lower left, showing a one-to-one correspondence between the two versions of the image. Pixels that were black in the original image are displayed as black, while pixels that were white are represented as white. But if you reverse the line, going from upper left to lower right, the pixels also become reversed. Black pixels in the original image display as white, and white pixels are displayed as black. The result is a negative.

By altering the shape and slope of the gray-map line, you can create interesting effects in an image. Increasing the slope of the line, for example, raises the level of contrast by forcing most of the pixels toward the light and dark ends of the gray-scale spectrum. You can adjust brightness by raising or lowering the line. If you reverse the gray-level for selected ranges of grays, you can get an effect called solarization in which certain areas appear as negative images and others as positives. By limiting the number of gray levels you can get an effect called posterization, in which the halftone appears more like a line drawing than a photo.

In most cases, you will not want to perform such radical transformations. But subtle changes in the gray-map editor can go far in improving the quality of your digital halftones. These changes can correct distortions in gray values that occur at the input and output stages of the halftone production process. Many gray-scale scanners darken gray values in the midtone areas of the image. For example, a scanner may reproduce a 50-percent gray value in the original photograph as 80-percent gray or more. On the output side, many imagesetters show little variation in gray levels at either end of the gray-scale spectrum. A 10-percent gray shade looks white, and an 80- or 90-percent gray shade looks black. Some image-editing programs have automatic calibration functions to compensate for these distortions. But you can use the gray-map editor to perform a similar function. By giving the gray map a convex (bow-shaped) curve, you can lighten the gray values in midtone areas. By setting the darkest shades (also known as shadows) to 90 percent and the lightest shades (known as highlights) to about 10 percent, you can increase the range of gray shades the image can display.

Another useful gray-scale editing tool is the histogram (Figure 6-15), a graph consisting of vertical lines that show how gray levels are distributed in an image. Each line represents a certain shade of gray, from zero percent

**Figure 6-15.** The histogram shows how gray shades are distributed in an image. Long lines indicates that a high percentage of dots have that particular gray shade.

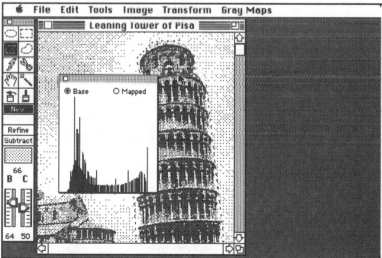

(white) on one side to 100 percent (black) on the other. Line lengths vary depending on how many pixels match the gray shade it represents. The best-looking images are generally those in which gray levels are distributed more or less evenly through the image. The number of light gray pixels, for example, should be roughly equal to the number of dark gray pixels. If the range of displayed gray levels is concentrated too much at the light or dark ends, detail can be lost.

The histogram itself does not allow you to alter gray values in an image. This is done with the gray-map editor or brightness and contrast controls. Many image-editing programs also include an Equalization function that explicitly redistributes gray shades in the image, spreading them more evenly across the graph. Be careful, however: gray values might be unevenly distributed due to the special nature of a particular image, and equalization could actually detract from image quality. Fortunately, most image-editing programs include an Undo function that allows you to return to the original distribution of gray shades.

Painting tools can also help enhance image quality by allowing you to fix scratches or perform other corrections. You can also use them to create original images and special effects. Unlike monochrome paint programs, which limit you to painting in black or white, image-editing programs allow you to paint in varying shades of gray selected from a palette or from the image itself. Cloning and texture tools allow you to click on a portion of an image and replicate it elsewhere. Click on a person's face, for example, and you can recreate the face—and everything around it—by painting in a different area of the image. Texture functions work in a similar manner, except they recreate the pattern immediately underneath the paintbrush rather than the area around it.

## Selection Tools

Almost all image-editing programs include functions that allow you to select a portion of an image on which you can

**Figure 6-16.**

Selection tools allow you to choose a portion of an image for further editing functions.

then perform a variety of operations (Figure 6-16). These typically include a marquee tool for selecting rectangular areas and a lasso tool for objects with irregular shapes. Once an object is selected, you can move it, duplicate it, or cut and paste it to another area or a different image file. Some programs allow you to stretch or rotate selected image portions. You can also limit certain software operations to the area inside or outside the selected area. This is known as masking.

Some image-editing programs include a helpful "Paste-If" feature that allows you to control how cut-and-paste operations are performed. This allows you to specify a range of gray values that will be pasted over when creating a composite photograph. For example, if you want to paste a photo of clouds onto a photo showing a clear sky, you can specify the relatively light gray values of the sky as the only area of the image to be pasted on. When the clouds are dropped into the image, they cover the sky only, leaving other areas with different gray values unobscured.

*Filters*

Image filters allow you to alter the characteristics of pixels in selected portions of an image (Figure 6-17). These can be

**Figure 6-17.** Image filters produce a variety of special effects, such as Blur (center), Sharpen (right), and Trace Edges (left).

used to create special effects and to compensate for irregularities. Blur and soften filters, for example, reduce the contrast in an image, making it appear slightly out of focus. They are useful if the background in a photograph is so sharp that it diverts attention from objects in the foreground. Sharpening filters increase the contrast around edges in selected areas. Diffusion filters rearrange gray levels in small portions of an image, creating a special effect known as a mezzotint.

Edge-tracing filters provide even sharper contrast, converting the image into bit-mapped line art. However, images created in this manner should not be confused with object-oriented graphics created by programs like Adobe Illustrator or Corel Draw. To convert a bit-mapped halftone into an object-oriented image, you need to use an auto-tracing program, such as Adobe Streamline.

## Using the Image

For many users, the final destination of a gray-scale image is a desktop publishing program. Almost all of these packages offer functions that allow you to crop or scale images. To scale an image means to reduce or enlarge it. To crop an image means to cut away a portion of it. This may be needed for aesthetic reasons if, for example, you want to draw attention to a particular object in a photograph. Cropping may also be required for design purposes. If you want to fit a horizontally oriented image into a vertical space, you'll need to crop away on the right or left. Of course, you can also do any necessary cropping or scaling when the image is scanned.

In addition to cropping and scaling features, some desktop publishing programs include image control functions that allow you to determine how scanned images are printed (Figure 6-18). These include brightness and contrast controls and simple gray-map editors. You can also set the dot shape, dot angle, and resolution of halftone screens.

The screen frequency you choose largely depends on your quality requirements and the type of paper stock on which you want to reproduce the image. If you plan to reproduce

**Figure 6-18.**
PageMaker and Ventura Publisher both include Image Control functions for gray-scale images.

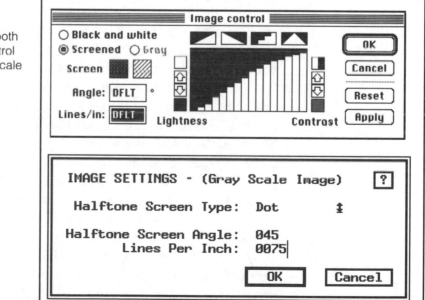

your pages on an absorbent paper stock like newsprint (the kind used in newspapers), maximum recommended screen frequency is 75 to 85 lpi. If you print on higher quality uncoated paper, maximum recommended screen frequency is about 100 lpi. Coated paper, used in most magazines, can handle halftones in the 100- to 150-lpi range.

Linotype engineers have found that certain screen frequencies produce the best halftone quality depending on the screen angle and imagesetter resolution. These recommended frequencies account for how the imagesetter builds the halftone cells. This doesn't mean that you'll get bad-looking halftones if you don't use these frequencies, but the recommendations do take full advantage of the imagesetter's imaging capabilities. Table 6-2 shows the recommended frequencies for dot angles of 45 degrees at 1270, 1693, 2540, and 3386 dpi on the Linotronic 330. For the optimum screen frequencies for other Linotronic imagesetters, call Linotype Co. or consult with your service bureau.

### Producing the Image

Assuming you have made any necessary modifications to the image, printing it on the Linotronic imagesetter is a relatively simple matter. One consideration is the output resolution to use. If you plan to produce an image with 256 levels of gray and a screen frequency of 100 lines or more, you probably need to go at the maximum resolution of 1693, 2540, or 3386 dpi depending on the imagesetter model you are using. To see why this is so, go back to our discussion above of the trade-off between resolution and gray levels. To compute the minimum resolution needed for various screen frequencies, take the square root of the number of gray levels in the image and multiply it by the desired screen frequency. In an image with 256 levels of gray, the square root is 16, which corresponds to the number of dots on each side of the halftone cell. Therefore, you need a resolution of at least 1600 dpi (16 x 100) if you want to produce a 100-line screen. If you want to produce a 150-line screen, you need a resolution of at least 2400 dpi (16 x 150).

**Table 6-2.**
Recommended screen frequencies for Linotronic 330 imagesetter, 45-degree screen angle.

| | 1270 dpi | 1693 dpi | 2540 dpi | 3386 dpi |
|---|---|---|---|---|
| **Under 70 lpi** | 64 | 63 | | |
| **70-79 lpi** | 75 | 75 | | |
| **80-89 lpi** | 90 | 86 | 86 | |
| **90-99 lpi** | 100 | 96 | 100 | |
| **100-109 lpi** | | | | 104 |
| **110-119 lpi** | 112 | 120 | 112 119 | 114 120 |
| **120-129 lpi** | | | 128 | |
| **130-139 lpi** | | | 138 | 133 |
| **Over 140 lpi** | 150 | | 150 | 141 |

Two other considerations when producing halftones on an imagesetter are the media—resin-coated paper vs. film— and the type of processor used to develop those media. Film generally produces superior results compared with paper. It tends to hold the halftone dots better, and it can save a generation in the printing process. You should also be sure to use a deep-bath processor when producing halftones of the imagesetter. These considerations are covered in greater detail in Chapter Nine.

In this chapter, we have discussed the issues involved in creating gray-scale halftones. For many users, this is the primary reason for choosing a Linotronic imagesetter over a 300-dpi laser printer or other low-resolution output devices. In the next chapter, we will discuss the much more complex subject of producing halftones in color. Much of the information in that chapter will build on our discussion of gray-scale images.

# COLOR INSERT

**Figure C-1**. Visible light includes all colors of the spectrum from red (left) to violet (right).

**Figure C-2.** The additive primaries, red, green, and blue (left), combine to make white. The subtractive primaries, cyan, yellow, and magenta (right), combine to make black.

**Figure C-3.** In process color reproduction, a photograph is separated into its four component colors.

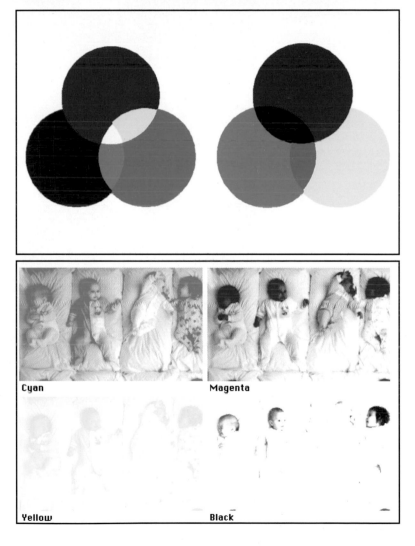

Cyan

Magenta

Yellow

Black

**Figure C-4.**
Producing good-looking color images can be a challenge. This image has an unrealistic red cast.

**Figure C-5.** An 8-bit image (left) does not look as realistic as a 32-bit image (right).

**Figure C-6.** Adobe Photoshop includes a number of filters (below) that can transform the appearance of an image. The three examples to the right show the effect of the Gaussian Blur (top), Mosaic (middle), and Sphereize (bottom) filters.

| ☐ Filter |
| --- |
| Add Noise... |
| Blur |
| Blur More |
| Custom... |
| Despeckle |
| Diffuse... |
| Facet |
| Find Edges |
| Fragment |
| Gaussian Blur... |
| High Pass... |
| Maximum... |
| Median... |
| Minimum... |
| Mosaic... |
| Motion Blur... |
| Offset... |
| Pinch... |
| Ripple... |
| Sharpen |
| Sharpen Edges |
| Sharpen More |
| Sphereize |
| Trace Contour... |
| Twirl... |
| Unsharp Mask... |
| Wave... |
| Zigzag... |

**Figure C-7.** Letraset's ColorStudio.

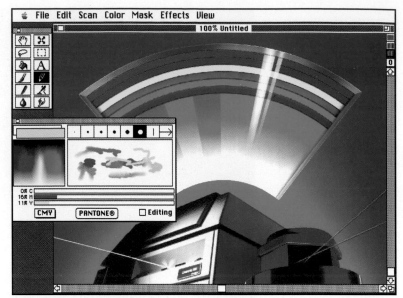

**Figure C-8.** Many desktop publishing programs let you select colors from the Pantone Matching System.

**Figure C-9.** Trapping corrects the gaps (left) produced by misregistration. Increase the line weight of the circle (center) and set its color to yellow (right).

# CHAPTER 7

# Producing Color on Linotronic Imagesetters

Not long ago, desktop publishing was largely a world of black and white. For many users, the ability to produce high-resolution text and graphics with inexpensive desktop computers—even in monochrome—was miracle enough. It was sort of like the early days of television when people marveled at watching their favorite shows from the comfort of their homes, despite the limited capabilities of early black-and-white TV sets. Now, of course, color rules the airwaves, and reruns in black and white seem like ancient relics. Some studios have even adopted the controversial practice of "colorizing" old movies and TV shows to increase their potential audience.

A similar phenomenon has hit the desktop publishing market. What began as a largely monochrome application has been smitten by the latest buzzword: "Desktop Color." With it, users have learned a new vocabulary that includes such terms as "Pantone Matching System," "trapping," and "CYMK." Countless vendors have released color-oriented hardware and software, and hardly a week passes by when a trade journal isn't touting the benefits—or enumerating the drawbacks—of some color publishing product.

It should be easy to see why people are so excited. Color has impact. It grabs attention. It establishes mood. Colors can be hot or cool, loud or muted, hard or soft. They can reflect the wholesomeness of a Norman Rockwell painting—or the tortured visions of Hieronymous Bosch. Fashion experts tell us to wear certain colors if we want to climb high on the career ladder. Companies spend thousands of dollars to find out which colors to use in their corporate logos—because they know the impact those colors will have on their public image.

Color is also complex. In previous chapters, we have seen the kinds of challenges posed by gray-scale images. Color increases those challenges exponentially. Almost anyone can spot a poorly rendered color image, but it takes skill and patience to produce a good-looking one, especially from the desktop.

Traditional color production requires enormously expensive pre-press equipment run by experienced operators. Hardware and software developers have made huge strides in bringing these capabilities to the level of the desktop, but users still need to educate themselves if they want to take advantage of these capabilities. And even experienced users will find themselves frustrated by the many limitations in current desktop color systems.

Linotronic imagesetters are a key component in the trend toward desktop color. With their high resolution, PostScript compatibility, and ability to work with film, they fulfill the basic requirements for color output. When we say "color output," of course, we don't mean that the imagesetter produces images in color. Instead, it produces the separations—one for each color on the page—that a commercial printer uses in color lithography.

In this chapter, we'll explore the many issues involved in producing color pages on Linotronic imagesetters. We'll begin with a general discussion of color perception, in real life and on the printed page. Then we'll discuss the traditional methods of color reproduction that desktop systems are trying to emulate. We'll go on to describe the hardware and software tools that make desktop color possible. Finally, we'll describe the process of producing color from the desktop, beginning with the initial scan and concluding with color separations from a Linotronic imagesetter. Along the way, we will learn to appreciate the complex interrelationships among products that handle color input, manipulation, display, and output.

We've included in this chapter a four-page color insert that demonstrates many of the concepts and products we discuss in this chapter. All of the images in this insert were produced using color desktop publishing software and a Linotronic imagesetter. Text references to figures in the color insert use the letter C rather than the chapter number.

Readers should keep in mind that this area of desktop publishing is evolving even more quickly than the industry as a whole. We'll try to stick to the basic concepts of color reproduction, but we cannot avoid discussions of specific products. These products are ever-changing, and new ones with more powerful capabilities are continually coming to market.

So let us begin by exploring the way human beings perceive color.

## Color Perception

Color begins with light. Visible light, to a physicist, consists of electromagnetic energy that falls within a certain range of wavelengths. To a child holding up a prism to a sunbeam, light is all the colors of the rainbow: red, orange, yellow, green, blue, indigo, and violet (to remember them, just think of a colorful fellow named Roy G. Biv). Our friend the physicist will tell us that each of those colors represents specific wavelengths within the overall spectrum of visible light (Figure C-1).

Look at a basket of fruit. What do you see? The apples are red, the bananas yellow, and the oranges, well, orange. But what we are really seeing is reflected light. Objects that appear to be a certain color do so by absorbing some of the colors in the spectrum while reflecting others. An orange, for example, absorbs all colors but orange, which is reflected back to the eye. An apple absorbs everything but red—unless it's the Golden Delicious variety, of course.

These reflected colors bounce off receptors in our eyes known as cones. Cones come in three varieties, one sensitive to red, one to blue, and one to green. By combining the input from the three kinds of cones, our brain gets its colorful view of the world.

## Color Models

The basic colors that can be combined to form other colors are called primaries. Red, green, and blue are known as additive primaries: add together equal percentages of red, blue, and green light and you get white (Figure C-2). Green and red in equal percentages produce yellow. Blue and green produce cyan; blue and red produce magenta. By varying the percentages of each additive primary, you can generate any color in the spectrum. Color viewed in this way is known as transmissive color, because it is produced by transmissions from luminous sources, such as TVs or computer monitors.

The colors produced by the union of the additive primaries—cyan, yellow, and magenta—are known as subtractive primaries. These are the colors used in printing inks. They are called subtractive primaries because they absorb all the colors of the spectrum except for those that are reflected back to the eye. Each subtractive primary represents two additive primaries—red, green, or blue—without the third. Again, the subtractive primaries can be combined in various percentages to produce almost any color. Because they are perceived as colors reflected back to the human eye, images produced on paper with these subtractive primaries are known as reflective copy.

RGB and CYM—Red, Green, Blue, and Cyan, Yellow, Magenta—represent two different color models, or ways of describing colors (Figure 7-1). CYM, as we have noted, is the model used in commercial printing and other applications where color is reflected back to the eye from a non-luminous source. The RGB model is used in scanners, displays, and other products where color is transmitted from a luminous source.

A third model, known as HSL for Hue, Saturation, and Lightness, is also used for color measurement. Two similar models are HSV, for Hue, Saturation, and Value, and HSB, for Hue, Saturation, and Brightness. Hue represents the color itself, from red through violet. Saturation represents the purity of the color. For example, if a color includes a high degree of gray, its saturation is low. Saturation is high when grays are minimized. Lightness or value represents the brightness of the color. Again, almost any color shade

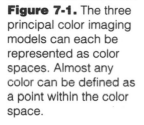

**Figure 7-1.** The three principal color imaging models can each be represented as color spaces. Almost any color can be defined as a point within the color space.

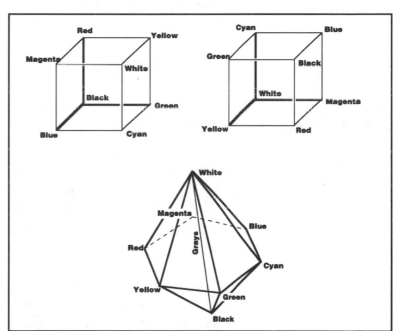

can be represented as HSL, HSV, or HSB values. Some people find that these are the most intuitive forms of color description.

### Effect of Lighting

We can see that light has much to do with our perception of color. A color photograph can look much different depending on whether you see it under candlelight in a darkened room or outside in bright sunlight. Graphics experts use the term "color temperature" to describe the degree of illumination in a particular environment. It is measured in degrees Kelvin. Standard daylight has a color temperature of about 5000 degrees Kelvin. An office illuminated by fluorescent light has a relatively cool color temperature of 4100 degrees. The standard Macintosh display has a color temperature of 9300 degrees. Some display products allow you to adjust the color temperature of a computer monitor so it can more accurately show color images.

Color perception is a complex process influenced by many variables. That's why it is such a challenge to produce good-looking color images, especially with desktop systems. Now we'll look at the two principal categories of color reproduction.

# Spot Color Reproduction

The simplest and least expensive form of color reproduction is spot color. As its name implies, this is a single color applied to one or more elements on the page. For example, an ad in the local Yellow Pages might have the shop's name and phone number in red, and the rest of the information in black.

Producing spot color is relatively easy. The lithographer makes a printing plate for each color to be used in the ad. For example, the elements to be printed in red appear on one plate. The items to be printed in blue appear on a second plate. Items to be printed in black go on a third plate. Pages are then run through the press three times, one for each plate and ink color.

## Preparing Copy

Camera-ready copy for spot color lithography can be prepared in one of two ways. The easiest method is to print the entire page in black and white, then use a tissue overlay to indicate the portions that should be printed in different colors. When making the spot color plate, the lithographer masks any area of the page that won't be printed in the extra color. The alternative is to produce three separate versions of the page (four if you add a second spot color). One, known as the composite page, includes all text and graphic elements to be printed. The second page includes only those elements to be printed in the spot color. The third page includes everything else. Each page, known more technically as an overlay, typically includes registration marks that allow for proper alignment.

## Pantone Matching System

One tricky aspect of working with spot colors is identifying them. Suppose you want to produce a flyer with a headline in a particular shade of blue. How do you communicate that request to the lithographer without submitting a sample? One answer is found in the Pantone® Matching System.

The Pantone Matching System is a set of standardized colors developed by a New Jersey-based company called Pantone Inc. Pantone publishes a catalog showing hundreds of colors used in commercial lithography. They begin with nine basic colors, including three shades of red and two shades of blue. These colors are identified by name: warm red, rubine red, reflex blue, and so on. Most of the remaining colors in the catalog are mixtures of these basic shades identified by a code number. For example, Pantone 151 is a mixture of 12 parts Pantone Yellow and four parts Pantone Warm Red that combine into a shade of orange.

Pantone licenses its color-coding scheme to manufacturers of printing inks. If you specify a Pantone code in your print job, a press operator can refer to the company's guide for the ink mixture needed to produce the specified color. The Pantone catalog also includes a selection of seven "Day-Glo" colors and seven metallic colors, including gold, bronze, copper, and silver.

Pantone color-matching guides are available at art supply stores for $100 to $150. Pantone publishes the guides on coated and uncoated stock to account for differences in how colors are seen on each type of paper. If you purchase one, be sure it is printed on the type of paper you intend to use for your color work. Pantone also recommends that the guide be replaced every year or so due to "uncontrollable pigment fading, varnish discoloration, and paper aging."

The Pantone Matching System is not the only way to specify spot colors. The four process colors, cyan, yellow, magenta, and black, can also be blended into spot colors. This is generally preferred when you have a lot of spot colors to apply, if your printer does not use the Pantone System, or if you just don't want to go to the expense of purchasing the color-matching guide. Another alternative to Pantone is to use a competing color-matching system. However, Pantone is the only color-matching system supported by desktop publishing products.

# Process Color Reproduction

Spot color reproduction is relatively straightforward and easy to master. Unfortunately, many printing jobs require a more demanding form of color reproduction known as process color. Process color is used to reproduce photographs and other complex images. It is also used as a substitute for the Pantone Matching System in jobs that involve spot color work.

In process color reproduction, images are broken down into four component colors: cyan, yellow, magenta, and black. We mentioned earlier that almost any color can be described as various shades of cyan, yellow, and magenta. However, these colors do not combine well when producing darker shades, especially black. For this reason, black is added as a fourth process color.

To produce a page using process color, the designer provides the lithographer with four pieces of film known as separations. The separations are printed in black and white, but each corresponds to one of the process colors: cyan, yellow, magenta, and black. The lithographer converts the separations into printing plates, and runs the pages through the press four times, once for each color. If everyone has done their job, the CYMK inks combine to faithfully reproduce the image (Figure C-3). In rare cases, a designer may add a fifth or sixth separation for an extra color that cannot be reproduced using the CYMK system. This is usually done when the job requires a metallic ink, such as gold or silver.

## Traditional Color Separations

Before the days of desktop publishing, separations for offset color lithography were produced largely by expensive pre-press systems (Figure 7-2) from companies like Crosfield, Scitex, and Hell (which is now a part of Linotype). Unlike microcomputer systems, which consist of standardized components from various vendors, color pre-press systems are proprietary. This means that all components are manufactured by a single company and sold as a package.

The heart of most color pre-press systems is a device known as a drum scanner. Unlike the scanners used in desktop publishing systems, a drum scanner serves as both an input and output device. True to its name, it includes a large drum on which color slides or prints are mounted, along with a second drum for output. As the first drum

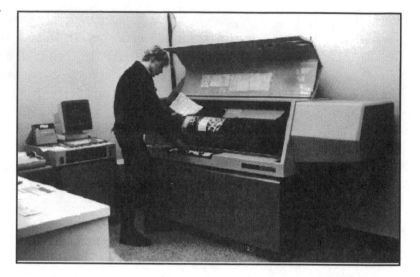

**Figure 7-2.** A traditional color pre-press system makes use of a drum scanner for producing color separations. (Photo courtesy of Wheeler/Hawkins)

spins, photosensitive cells with color filters convert the slide or print into a series of electrical signals. The scanner uses these signals to produce the separations on the second drum.

As an image is converted into separations, the scanner operator can make a number of adjustments to improve its ultimate appearance. Simple adjustments include cropping and resizing, but an experienced operator can also sharpen blurry edges or make subtle adjustments in the percentages of each color. Because of imperfections and inconsistencies in color inks, some CYMK combinations do not produce an accurate reproduction of the original colors. By adjusting the color percentages, the scanner operator can correct these kinds of problems. This kind of color correction requires a lot of skill and experience.

## Color Proofing

One important aspect of color reproduction is proofing. Color printing is tricky under the best of circumstances, and many designers want to see their pages before paying thousands of dollars for a commercial lithographer. Unfortunately, looking at four pieces of film is not going to give you a great idea of how your color image will look when it's printed. Because of this, several companies have devised methods for producing color proofs that provide a more-or-less accurate reproduction of the image. These systems include DuPont's Cromalin® and 3M's MatchPrint® and Color Keys (Figure 7-3).

In addition to these proofing mechanisms, color designers and lithographers use devices known as densitometers and colorimeters to measure printed color. A densitometer, true to its name, measures the density of a color or gray shade. However, it does so in negative terms. For example, if the densitometer measures the density of blue as zero, 100 percent of the blue light is being reflected or transmitted back to the eye. Densitometers, unfortunately, are not always accurate in measuring colors as they are perceived by the eye. Colorimeters, a similar type of device, are more accurate in measuring color as it will be perceived.

## Desktop Color

Over the past few years, many hardware and software developers have introduced products that allow production of color images using desktop computers. At first, most of

**Figure 7-3.** 3M's Digital Matchprint color proofing system.

these were oriented toward spot color. But many products now offer process color capabilities as well.

A major debate is raging over the merits of desktop color systems in comparison with traditional prepress systems. On the plus side, desktop systems are much less expensive than traditional prepress equipment, even if you include the cost of an imagesetter. Traditional systems are priced as high as $1 million. A PostScript imagesetter for producing color separations can be had for well under $100,000. A Macintosh II system with large-screen color display, 120-megabyte hard drive, eight megabytes of memory, and a full complement of software costs less than $20,000. Most desktop color publishers get along without their own imagesetters; they use a service bureau instead.

Desktop color also offers the best of two worlds. Programs like Aldus PageMaker and Xerox Ventura Publisher brought publishing capabilities to the masses by offering page production tools in easy-to-use packages. For the first time, users had complete control over the appearance of pages without the intervention of a graphic artist. Desktop color promises the same kind of freedom for color publishers.

### Limitations of Desktop Color

Unfortunately, microcomputer-based color systems have their limitations, especially in the areas of performance and output quality. One problem is that color images consume large amounts of memory and disk space—a slide scanned at 4096-line resolution and 24 bits of color requires about 60 megabytes. This can strain the capabilities of even the most powerful microcomputers. PC-based color systems may be cheap compared to traditional systems, but they still require a lot of expensive horsepower to handle color data. A large screen, 32-bit color display costs about $5000—and that doesn't include an accelerator or calibrator.

Another problem with desktop color systems is a lack of reliable proofing mechanisms. That $5000 color monitor can display a photograph, but the methods for generating an image on an electronic display—which uses additive colors—are much different than those for printing subtractive process colors. An image that looks good on the screen can appear to have distorted colors when printed. Color PostScript printers are even less reliable. Some service bureaus that do desktop color work offer traditional color proofing systems like 3M's Color Keys or MatchPrints. These are much more reliable than other proofing methods, but are expensive and unwieldy.

Finally, there's the question of quality. Most publishers of magazines and other color publications have high quality requirements, and many believe that desktop systems cannot meet their demanding standards. Though a few magazine publishers produce their color pages with desktop software, the vast majority still rely on traditional color separation systems.

Does this make desktop color systems mere curiosities? Not at all. Just as desktop publishing systems brought typesetting capabilities to the masses, desktop color systems bring color capabilities to publishers previously limited to a world of black and white. Candidates for desktop color systems include small newspapers, catalog publishers, and packaging designers—users who were scared away from traditional color because of its cost and complexity.

The nice thing about desktop color is that the products are getting better all the time. Many desktop publishing programs can now produce process color separations, even of photographs. Color graphics packages allow users to manipulate and enhance color images. Color display products are getting faster and more reliable.

## Standards

One distinction between desktop systems and traditional pre-press is the use of standardized components. Traditional pre-press systems, we have noted, are based on proprietary equipment that cannot be used in systems from other vendors. Desktop users, however, can purchase a computer, display, printer, software, and other products from different vendors and configure their own systems. Almost all output is produced using PostScript, the *de facto* standard for page description languages. Standard file formats like TIFF and EPS can be used with a wide range of programs in multiple hardware environments.

This ability to mix and match system components is one reason why desktop color is so much cheaper than traditional color. Vendors of proprietary systems essentially have a captive audience that must purchase all components from a single source, and they charge accordingly. In an open system, many different vendors compete to supply the system components. This, and the larger market for microcomputer products, results in lower prices.

But there is also a downside to open systems. In a proprietary system, the manufacturer can be sure that all components work smoothly together to produce consistent, high-quality output. But when systems include products from multiple vendors, achieving this consistency proves to be more difficult. A slide captured on one company's scanner might look different than the same image captured with a competing product. A photograph might appear to have too much red on a certain monitor when the problem is really in the display, not the picture. Images that appear too light on one imagesetter might be too dark on another (Figure C-4).

Hardware and software developers are addressing these problems with new products that allow users to compensate for differences among the various components in a system. As the standards that govern desktop color production are refined and extended, hardware and software products will no doubt learn to better cooperate with one another.

The most important standard in the desktop color market is PostScript. As PostScript devices, Linotronic imagesetters are vital components in color production. They perform the ultimate step in the process: producing the same kinds of color separations that a drum scanner would produce in a traditional pre-press system. But users must be sure they have the right machine for the job. Early imagesetters—and their RIPs—were designed for monochrome output, and proved to have limitations as color machines. Newer models, especially the Linotronic 230, 330, and 530, have improved color output features.

One challenge for the color imagesetter is producing separations with a high degree of repeatability. This means that dots are correctly aligned from one separation to another. If the separations are out of registration, the CYMK colors will not mix properly and the image will probably be distorted. Poor repeatability can also cause problems with trapping, a spot color technique described later in this chapter. Repeatability is measured in microns. Dot repeatability of two one-thousandths of an inch or less is considered sufficient for color production. The key to improved repeatability in the Linotronic 230, 330, and 530 is a patented mechanism for feeding film separations through the imagesetter.

## The Tools of Desktop Color

Before we get to the imagesetter, we need to produce our color pages with a desktop publishing system. It should be apparent by now that the cute little PC that sits on your desk for simple word processing applications may not be sufficient for color work. It takes a powerful combination of hardware and software tools to meet the requirements for color production.

### Computer Workstation

First, there is the computer itself. Most of the exciting new products for color production run on the Macintosh, though IBM-compatible computers are beginning to catch up. The bottom line is that you need plenty of horsepower, especially when doing process color work. It takes a lot of data to describe a color image, so the computer must be fast and able to work with large amounts of memory. It should also be expandable, since you may need to add a display controller, scanner board, or other add-on products.

In the Macintosh environment, almost all color work is done on the Macintosh II series; earlier models are limited to monochrome display. In the IBM environment, computers based on the 80286, 80386, or 80486 microprocessors are preferred. The 386 and 486 chips are best, but 486 systems tend to be quite expensive.

Whatever type of computer you use, you'll need plenty of memory to work with color images. The minimum configuration for the Macintosh is four megabytes, with eight megabytes being preferred for process color work. PC users can get by with two megabytes, but again, the more memory the better. As more sophisticated color software moves to the PC, memory requirements will no doubt rise.

*File Storage*
As we have seen, color images consume large amounts of disk space. One manufacturer of a color scanner recommends that users purchase a 600-megabyte hard drive to store scanned images. It may sound like a lot, but one slide scanned at maximum resolution can consume up to 60 megabytes. You don't have to go as far as ordering a 600-megabyte disk drive, but you do need sufficient storage space to work with color images. Depending on the kind of color work you plan to do, this could mean anything from a 65-megabyte drive on up.

For many users, the main storage medium—the hard disk—is less important than the media used to transport color images to a service bureau. Relatively small images can be stored on 3.5-inch diskettes with capacities ranging up to 1.44 megabytes. If an image file is too large to fit on a single diskette, the user can use a backup program to divide the file among multiple disks. File compression programs available in many public domain software libraries can also reduce file size.

In some cases, standard 3.5-inch diskettes are not enough. Many users working with especially large image files find themselves compelled to purchase removable hard drives. These products, from companies like Iomega and SyQuest, allow storage of large amounts of data—typically 20 to 40 megabytes—on removable disks. A user can easily copy all or most of the files involved in a color job and transport the disk to the service bureau. Unfortunately, there are few standards in the market for high-capacity removable disks. An Iomega disk, for example, will not work with a SyQuest drive. This means that before purchasing such a product, you should check with your service bureau to be sure they have a compatible drive. Other alternatives for transporting your files—as you'll see in Chapter Eight—are portable drives and erasable optical drives.

Some companies have developed file compression products that can vastly reduce the size of color image files, making them easier to store and transmit. These products are sold in the form of software-only packages and hardware/software combinations. The latter are more expensive, but offer faster compression and decompression speeds.

*Color Displays*
The display is one of the most important elements in a desktop color system, especially given the absence of reliable proofing mechanisms. A good color display gives the user a reasonable idea of how an image will look when it is produced. Working with a bad display is like trying to paint a picture in the dark.

The primary distinction among color displays is the number of colors they can show for each dot (generally known as a "pixel") on the screen. This number is generally referred to in terms of "bits per pixel." A bit is the primary unit of data in a computer system. It can have one of two values: zero or one. A number described with eight bits of data can have any one of 256 values. A number described with 24 bits of data can have any one of 16.8 million values. Each of these values can be used to represent a gray or color tint. An eight-bit image contains 256 colors or gray shades; a 24-bit image contains 16.8 million colors.

Many color monitors are limited to displaying eight bits, or 256 individual colors. This may sound like a lot, but images viewed on an eight-bit display tend to have a grainy, unrealistic appearance. Costlier monitors have 32-bit capability. In this case, the display can show 8 bits, or 256 shades, for each of the three primary additive colors—red, green, and blue—with 8 bits left over for other data. These monitors can show images that appear as realistic as a color photograph (Figure C-5). Theoretically, a 32-bit display can show 16.8 million colors. In reality, few monitors can actually display that many colors because they don't have 16.8 million pixels on the screen.

All displays, color or monochrome, conform to certain standards. On the Macintosh, that standard is known as QuickDraw, which is Apple's name for the software that allows the computer to show images on the screen. Most color QuickDraw displays use 8 bits of color, but 32-bit QuickDraw is used in many large-screen color monitors.

The most popular color display standards in the IBM-compatible environment are CGA, EGA, and VGA. VGA, the most recent standard, is preferred for most color production work, especially with photographs. Standard VGA offers a choice of 320 by 200 pixel resolution and 64 colors, or 640 by 480 and 16 colors. Users can boost resolution and bit depth by using Super VGA displays or the 8514/A extension for VGA. CGA, the oldest standard, is seriously limited in its resolution and the number of colors that can be displayed, and should be avoided.

Several companies manufacture products that enhance the use of color displays for publishing work. These include graphics accelerators and calibrators. Graphics accelerators are boards that speed up the display of color images, and are especially useful with 24- and 32-bit monitors. Calibrators are hardware/software products that adjust the way a monitor projects colors to account for lighting conditions and changes in the monitor itself.

One such product, the PrecisionColor Calibrator from Radius, includes a sensing device that attaches to the monitor along with utility software that performs the calibration (Figure 7-4). In addition to altering the mix of RGB color components, the calibration system can adjust the display's color temperature to simulate various proofing situations. For example, you can set the temperature at 5000 degrees, which is the industry standard for proofing color prints.

Another feature is the ability to account for the monitor's gamma distortion. A gamma curve is a graph that com-

**Figure 7-4.** The PrecisionColor Calibrator from Radius allows you to adjust the display to show colors more accurately.

pares the brightness of an image stored in the computer with the same image as it is displayed. Ideally, the graph should show a diagonal line representing a one-to-one correspondence: each pixel in the displayed image should have the same brightness as the stored image. But most displays have a slightly distorted gamma curve, which tends to make images look slightly darker or lighter than they should. The PrecisionColor calibrator can adjust the curve to ensure accurate image display.

Some software products offer a crude but inexpensive form of display calibration that depends on the human eye. In Letraset's ColorStudio, for example, the manual includes a color photograph of men in a sailboat. When the program is launched, it displays the photograph on the computer screen. A user can then adjust the monitor's brightness and contrast to get the closest possible match between the printed and displayed versions of the photo. Other programs use a color ramp—a bar displaying a wide range of colors—in place of a photograph.

### Scanners

The scanner is another key component in a desktop color system. Like their black-and-white siblings, these products represent a bridge between real-world images and the digital environment of the page layout. With the proper scanner, you can incorporate color photographs and other artwork into your publications and ultimately have them printed as separations.

All color scanners have one thing in common: they capture images as combinations of red, green, and blue. Depending on their capabilities, they can sense anywhere from 4 to 12 bits of information for each of these colors. Some scanners capture the red, green, and blue elements in three separate passes. Others can capture all three in a single pass.

Color scanners used in desktop publishing systems fall into

four categories: hand-held, flatbed, slide, and drum. Some video digitizers also offer color capability.

Hand-held scanners, the least expensive option, are priced at less than $1000. They can be useful for scanning position-only images, but should be avoided for final production work. Resolution is often limited to less than 100 dpi. Pixel depth—the amount of information that can be stored for each dot in the image—is also limited. Some hand-held color scanners, for example, can sense just 16 gradations for each of the three primary colors. This compares with 256 gradations for other types of scanners.

Flatbed scanners, priced between $2500 and $10,000, offer resolutions ranging from 300 to 800 dpi, with most providing 300 to 400 dpi. In most models, a sensing device passes over the image three times, one each for red, green, and blue. They can theoretically sense up to 256 shades of each color, but distortions in the scanning process can reduce this to between 64 and 128 shades. Images captured by flatbed scanners can look washed out compared with traditional pre-press output. But considering the price-to-quality ratio, some publishers find that flatbed scanners are sufficient for their needs.

Slide scanners, priced between $8000 and $20,000, are the best choice for professional users who produce color separations of photographs. True to their name, they capture images stored on slides and transparencies, and produce images superior to what a flatbed scanner can produce from a color print. Resolutions range from about 1000 lines per slide to 4096. Most slide scanners can sense 8 to 12 bits of information per color per pixel. Distortions in the scanning process reduce this somewhat, but slide scanners can still sense millions of different colors. Leading models include the Color Imaging System from BarneyScan Corp., LS-3500 from Nikon, and Array Scanner One from Array Technologies. The latter product can also be used as a camera in a video digitizing system.

One product that occupies an area between flatbed and slide scanners is Truvel's TZ-3. This scanner, which includes color as an option, uses a camera-like mechanism that sits over a flatbed to capture images. Because of its design, it can capture three-dimensional objects in addition to images on paper. By using a zoom feature, users can increase the effective resolution to 900 dpi.

At the top of the price scale are digital drum scanners. These products use the same technology as the drum scanners found in traditional pre-press systems. But instead of passing the image directly to the recording drum, the digital drum scanner stores the data in a digital format that can be imported into a computer system. They typically cost $50,000 or more.

*Proofing Systems*

Proofing represents one of the most difficult aspects of desktop color. Aside from the color display, users have little opportunity to see how their pages will look when printed. One product that does allow a form of proofing is the color PostScript printer. Almost all color separations produced by desktop publishing software are stored as PostScript files. Because of this, any PostScript device can provide a preview of the image—up to a point.

The first color PostScript printer on the market was the QMS ColorScript 100, released in 1988 (Figure 7-4). Since then, several manufacturers have released their own color PostScript printers. In general, these products offer 300-dpi resolution and are priced between $7000 and $12,000. Unlike most other PostScript printers, they do not use a laser beam to create an image of the page. Instead, a heating element melts dots on to the page from plastic or wax-covered sheets corresponding to the three CYM colors. Most color PostScript printers are thus known as thermal-transfer devices.

Color printers work best for previewing spot color work or

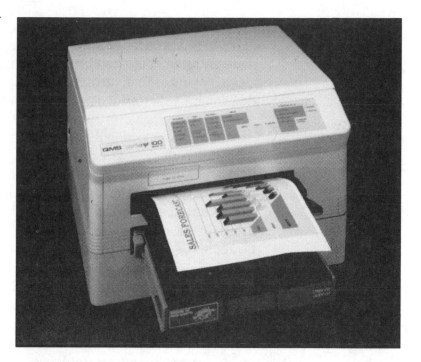

**Figure 7-4.** The QMS ColorScript is a popular color PostScript printer.

process color that does not include photographs. However, their 300-dpi resolution is generally insufficient for quality photographic reproduction. Users can get a rough idea of how the photo will look, but many subtleties in the image are lost. Another drawback is their high expense in comparison with monochrome PostScript printers. Many service bureaus, however, offer output on color PostScript printers as an adjunct to their imagesetting services.

A more reliable proofing alternative is provided by traditional proofing systems, such as 3M's Color Keys and MatchPrints. These systems use a variety of methods to produce realistic proofs from color separations generated on the imagesetter. However, they are expensive and unwieldy. The Color Keys system, for example, costs about $75,000. It uses heat lamination technology to produce the proofs, consuming large amounts of electricity and requiring more than an hour to warm up. Another disadvantage is that you cannot tell how your image will look until it has already been produced as separations on the imagesetter.

*Software*

Hardware, of course, only tells half the story when it comes to desktop color. New software packages have been essential to advances in digital color applications. In some cases, developers have added color production capabilities to existing programs. Other developers have introduced products that specifically address one or more aspects of color input, design, or production.

Software used in color production falls into many categories. Scanner software is used with the scanner to capture images and store them in one of several standard file formats. Color editing software allows the user to retouch, combine, and adjust color images to improve their appearance or create special effects (Figure C-6). Color illustration software uses an object-oriented approach to color design as opposed to the bit-mapped approach in color editing programs. Color-oriented desktop publishing software can incorporate color images produced with other programs, as well as applying colors to text and graphics created within the package. Color separation software is used to produce process color separations from files created by desktop publishing or graphics programs. Many software packages combine these functions. PhotoMac, for example, integrates color editing and separation features. Adobe Photoshop, Letraset's ColorStudio, and BarneyScan XP all combine scanner control and image-editing functions.

# Working With Color Scanners

Unless you plan to create an image from scratch, the color production process begins with scanning. All color scanners sold these days include some kind of software to control image capture (Figure 7-6). In some cases, the scanning software is a stand-alone program that performs the scan and little else. In other cases, scanning functions are built into a color image-editing package.

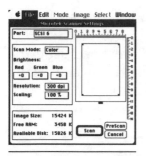

**Figure 7-6.** Scanner control programs sold with color scanners allow you to choose various image-capture settings.

Scanner control programs share several attributes. Most of them allow the user to perform a preview scan in black and white, at low resolution, or both. You can then select the portion of the image you intend to use. This feature is helpful given the enormous amount of data needed to store color images. If you only need half the image, you can reduce file size by 50 percent. Once the desired image area is selected, the software performs a color scan at full resolution.

The software should also allow you to resize an image as it is scanned. If you intend to produce an image that's three inches wide, it doesn't make a lot of sense to scan it at full size. Again, by reducing the image size, you can save valuable file space.

### Adjusting Resolution

Most scanner control programs allow you to adjust the resolution of the scan. In most cases, you should follow the same guidelines used for scanning black-and-white photographs: compute the amount of data required by the imagesetter to produce the image at the desired size, then set the resolution accordingly.

Some scanner control programs allow you to adjust color balance or other image attributes. For example, you may be able to change the mix of red, green, and blue primaries. The Nikon LS-3500 slide scanner includes software that can account for lighting conditions and the type of film used for the slide.

### Scanned File Formats

Another important feature in scanning software is the range of graphic file formats it supports. If your software supports a wide range of formats, you will have a lot of options for how you can use the image. Most of the formats used to store gray-scale images, including TIFF, PICT2, and EPSF, can also store color images. A fourth format, Targa, is a popular color format for video digitizers, especially in the IBM environment. TIFF is generally the preferred format for storing scanned images. EPSF is not recommended as an intermediate file format because of the large amount of file space it consumes. However, it is the preferred output format for color separations.

# Working With Desktop Color Software

Producing desktop color involves more than a single process and a single software package. In this section, we'll look at several different types of software packages that can produce color images on a Linotronic imagesetter.

### Color Image Editing Software

Once an image is scanned, its next destination is usually a color image-editing package, such as Adobe Photoshop, Letraset's ColorStudio (Figure C-7), or Avalon Development's PhotoMac. These programs allow you to perform a wide range of editing, retouching, and image manipulation functions. In many ways, they are similar to gray-scale image editing programs like Letraset's ImageStudio and Silicon Beach Software's Digital Darkroom. However, instead of manipulating gray shades, these programs allow you to work with each of the primary colors in the image. Although each package has its own unique features, a number of general functions are common to all.

*Painting Tools*
Painting tools in color image-editing programs are similar to those found in their gray-scale counterparts. A paintbrush tool is used to retouch portions of an image or to add special effects. A paint bucket tool, as its name implies, allows you to pour a color or pattern into a section of an image. Spray paint tools provide an airbrush effect. Other common tools include a "pencil" for adding and deleting individual dots; a "teardrop" for softening edges; and a "fingerpaint" tool that allows you to smudge colors in the image.

Most image-editing programs provide a high degree of control over how these tools are employed. For example, you can select a color from a portion of an image by clicking on it. Cloning functions, like those in gray-scale image-editing programs, allow you to click on a portion of an image and replicate it elsewhere. Texture functions work in a similar manner, except they recreate the pattern immediately underneath the paintbrush rather than the area around it.

*Selection Tools*
Selection tools allow you to select a portion of an image on which you can then perform a variety of operations. These typically include a marquee tool for selecting rectangular areas and a lasso tool for objects with irregular shapes. More advanced selection tools, such as the magic wand in Adobe Photoshop, allow you to click inside an irregular object to select it automatically. The program performs this feat by seeking pixels with similar color values, so it works best when there is a strong contrast between the object and surrounding areas.

Once an object is selected, you can move it, duplicate it, or cut-and-paste it to another area or a different image file. Some programs allow you to stretch or rotate selected image portions. You can also limit certain software operations to the area inside or outside the selected area. This is known as masking.

In addition to its standard selection tools, Letraset's ColorStudio has an especially powerful masking function. The program uses two "layers"—a color layer for image display and a masking layer that acts as a stencil. Most functions that can be used in the color layer can also be used in the masking layer. For example, you cut-and-paste the image of a tree onto the mask to create a cookie cutter effect. Any paint applied to the mask only shows through where the tree has been pasted. By adding gray shades to the masking layer, you can create a screen that filters out some of the paint, producing a tint effect.

## Image Filters

Filters are special effects that manipulate the color values in an image. In most cases, you can perform the filter operation on the entire image or selected portions. Common filters include blurring, sharpening, edge tracing, and despeckling, which removes random pixels from the image. Sometimes these are used to create special effects. In other cases they are used to improve the general appearance of the image.

## Composite Images

Using the program's cut-and-paste features, you can create images composed of sections from other images. For example, you may want to add clouds to a blue sky in a landscape scene. To do this, you need two images, one of the clouds and one of the landscape. Using selection tools, you can select the clouds from the first image, copy them to the program's clipboard, then paste them into the second image. Most image-editing programs provide some sort of blending function so that the first image blends smoothly into the second one.

## Color Adjustment

Color adjustment functions allow you to change the mix or intensity of colors or replace one color with another (Figure 7-7). There are several reasons why you may want to do this: to improve the general appearance of an image, to

**Figure 7-7.** Color image-editing programs, such as Adobe Photoshop, allow you to control color balance by varying the intensity of the component colors.

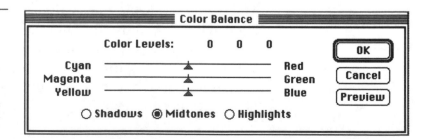

create certain kinds of special effects, or for calibration purposes.

In Chapter Six, we discussed a function in many image-editing programs known as the gray-map editor. Color mapping functions work in a similar manner, except that instead of working with shades of gray, we are working with shades of red, green, and blue, and ultimately the CYMK colors as well.

In some programs this is handled by means of a graph, similar to the gray-map editor, that shows the relationship between pixels in the original image and pixels in the display. The difference is that there is one graph for each color, plus a fourth graph that represents the entire image. By adjusting the transfer lines on these graphs, you can brighten or darken an image, or increase or decrease its contrast. A slightly curved transfer line can create a smoother transition among colors than a straight diagonal line. A concave transfer line—one that dips down in the middle—emphasizes the light (highlight) and dark (shadow) areas of the image while de-emphasizing the mid-tones. A convex line—one that curves up in the middle—emphasizes the mid-tones at the expense of highlights and shadows.

In addition to its remapping functions, Adobe Photoshop can adjust color balance by means of sliding controls for each color. You can choose to add or subtract the amounts of cyan, magenta, or blue in the shadow, midtone, or highlight regions of the image. You can also adjust hue and saturation using similar controls.

*Color Separations*
Finally, many color editing programs can produce CYMK color separations. Simply stated, this means creating four PostScript files—one each for cyan, yellow, magenta, and black—and having them produced on film media on a Linotronic imagesetter. But the reality is far from simple, as the software must often make adjustments to the image

file to ensure the best possible output results. Some software developers have automated this process by adding what are known as "lookup tables" for each output device to their programs. The lookup table contains information about the output characteristics of popular printers and imagesetters. The software uses the lookup table to make minor modifications to the image that will improve its appearance on that particular output device.

## Color Illustration Software

Illustration programs differ from color image-editing programs in that they work in an object-oriented format. Color images produced by a scanner or paint program are known as bit maps, because they consist of individual dots. Illustration programs, on the other hand, create images in the form of objects: lines, curves, circles, rectangles, and so on. On the computer screen, these objects appear to be no different from standard bit-mapped images. But the software sees these objects as mathematical expressions. One way to see the difference is to consider a simple image, such as a stop sign. In a bit-mapped format, the stop sign would be a simple array of dots organized into an octagon. In an object-oriented format, however, the stop sign would be stored as a series of mathematical instructions: go right for six inches, then at a 45 degree angle for 6 inches, then another 45 degree angle for 6 inches, and so on.

Object-oriented images offer a number of advantages over bit-mapped images. Because they are described in mathematical terms, the software can print them at the maximum resolution of the output device. Object-oriented images can be enlarged or reduced without loss of image quality. Finally, the objects that make up the image can be moved and manipulated with relative ease without affecting other objects.

Many artists prefer object-oriented illustration packages because they offer a high degree of precision over the drawing process. The leading packages on the Macintosh are Adobe Illustrator and Aldus Freehand. The leading packages in the IBM environment are Corel Draw, Micrografx Designer, Arts and Letters, and GEM Artline.

All of the illustration packages mentioned above can be used to produce color separations, usually for spot colors. Spot colors can be produced in one of two ways. You can produce separations for each of the spot colors using the Pantone Matching Guide, or produce process color separations that create spot colors as combinations of cyan, yellow, and magenta. Some of these packages allow you to create custom colors by mixing various percentages of red, green, and blue or cyan, yellow, and magenta. Once the illustration is created, it can be exported to a desktop publishing program in EPSF format or produced directly in the form of separations.

## Color Desktop Publishing Software

Many desktop publishing packages, including Aldus Page-Maker, QuarkXPress, and DesignStudio, have built-in functions for producing spot or process color. In most cases, colors can be applied to headlines, rules, boxes, and other page elements. They can be specified as combinations of process colors, or as colors from the Pantone Matching System (Figure C-8). You can also import color images from illustration or image-editing programs.

Your ability to produce color depends on whether you want spot color separations or separations of color photographs. Remember that spot color can be produced in one of two ways: as a separate overlay for each of the spot colors, or as process color separations that create spot colors as combinations of cyan, yellow, magenta, and black. PageMaker, QuarkXPress, and other desktop publishing packages can produce process color separations directly, but only if the process colors are used to create spot colors (Figure 7-8). If you want to produce process separations of color photographs or other TIFF files, you must use an add-on utility program that works with your desktop publishing software.

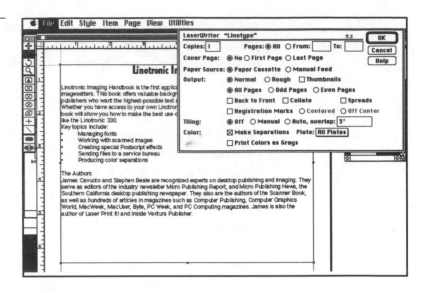

The leading color separation programs include Aldus
PrePrint (Figure 7-9), which works with PageMaker;
Letraset's DesignStudio Separator, which works with
DesignStudio; and Spectre Seps QX, a separation package
from Pre-Press Technologies that works with QuarkXPress.
Ozette Technologies offers a program, Color Sep/PC, for
producing color separations with Ventura Publisher and
the Windows version of PageMaker, but it cannot produce
separations of photographs. Adobe sells a program called
Adobe Separator that works with Illustrator files.

These packages work with files that have already been
created by the publishing software. With PrePrint, for
example, the process begins with a PageMaker document
that includes color TIFF images. The first step is to save the
publication in a format known as OPI, short for "Open
Prepress Interface." This is a special kind of file designed
for use with color separation programs. Using PrePrint,
you open the OPI file and produce the separations. These
separations can be printed directly to the output device, or
produced as EPSF files. PrePrint also includes functions
that allow you to enhance or adjust the image.

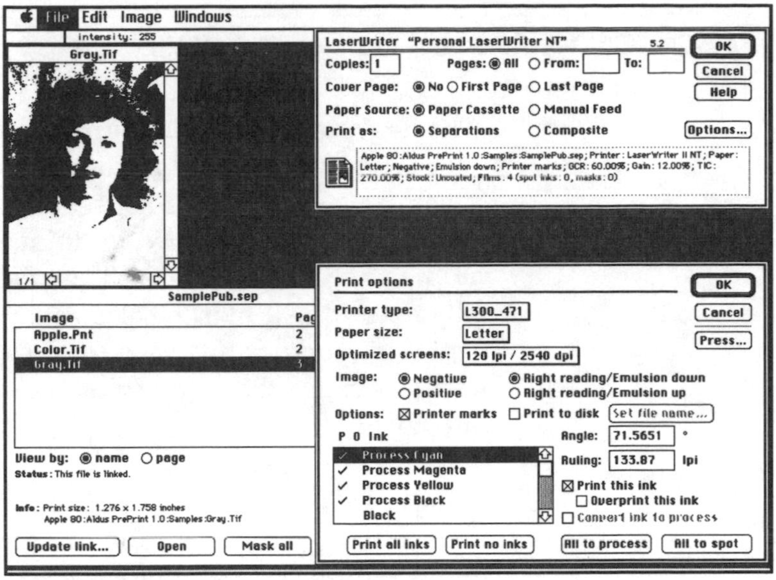

**Figure 7-9.** Aldus PrePrint produces process color separations from PageMaker files.

## General-Purpose Color Separation

The programs described above are designed for use with specific publishing or graphics programs. However, you can also purchase separation software, such as Avalon Development's PhotoMac, that works with a variety of packages, including PageMaker and QuarkXPress. PhotoMac creates color separations in one of two ways. The first method, used with QuarkXPress, separates an image into five EPSF files, four for actual output and one for screen display. The fifth file is placed in a QuarkXPress document at the desired location. You then use Quark's output function to produce four separation films.

Users of PageMaker and other publishing packages must install an output driver called PageSaver to produce separations. First you export the image from PhotoMac as an 8-bit PICT file for position only. After the file is positioned on the page, you print the publication to disk using the PageSaver driver. You then open the PageSaver file with PhotoMac to produce the separations.

# Producing Color on the Linotronic Imagesetter

The final step in the color publishing process is to produce four separations, one each for cyan, yellow, magenta, and black, on a Linotronic imagesetter. Each separation will include a halftone of the color image in which the dots are printed at different angles. The angle of the dots is generally determined by the color separation software. If the wrong dot angles are used, moire patterns are likely to result because the primary colors will not blend together properly.

Before you go to the time and expense of producing color separations, it is important that you communicate with representatives from your service bureau and print shop. These people can inform you of many factors you must take into account to produce the best possible output.

### Registration Factors

These factors can get pretty complex. For example, even the best printing press will have difficulty producing pages with perfect registration. In other words, the colors can shift slightly from one separation to the next. One way to compensate for this, at least in spot color production, is by using a method called trapping (Figure C-9). There are two types of traps: spreads and chokes. In a spread, areas printed in a light color are permitted to spread into areas printed in black. This eliminates any gap that might appear between the two areas. In a choke, dark areas are permitted to spread slightly into light areas, again eliminating any gaps. Some color separation packages allow you to determine the degree of trapping when you produce the CYMK overlays. Your print shop representative can give you a good idea of how much trapping may be needed. This form of trapping should not be confused with "ink trapping," an unrelated printing phenomenon in which inks mix improperly and produce distorted colors.

## Dot Gain

Another problem that occurs on the printing press is dot gain. When ink is laid on the paper, it tends to spread slightly, which results in an enlarged dot. The degree of dot gain depends on the type of ink and paper used in the printing process. For example, uncoated paper stocks generally produce more dot gain than coated stocks. Some color editing programs, such as Adobe Photoshop, offer features that allow you to compensate for dot gain. Again, it is important to communicate with your print shop to find out what kind of dot gain you can expect and how you might compensate for it. In general, if the degree of dot gain is high, the density of the dots on the separations should be reduced.

## Undercolor Removal

In some cases, it may be advisable to perform an operation known as undercolor removal to adjust the mixture of CYMK primaries on the printed page. This technique takes advantage of the black color component to reduce the total amount of ink needed to print a page. We mentioned earlier that a mixture of 100-percent cyan, yellow, and magenta produces solid black. Equal mixtures of the three primaries in lighter densities produce gray. Undercolor removal is the process of replacing "neutral color areas"—those with equal percentages of cyan, yellow, and magenta—with the appropriate shade of gray. A similar technique known as gray component replacement uses black ink to adjust the overall density of the image.

Some color-oriented graphics programs offer a form of undercover removal as part of their output functions. For example, in Adobe Photoshop, you can specify one of five degrees of black replacement—none, light, medium, heavy, and maximum—or a custom amount. To use this feature correctly, it is important to communicate with your printer to learn about the density of the inks they use on press.

In addition to reducing the amount of ink used on press,

undercolor removal also ensures greater consistency if you have to produce the same publication on multiple printing presses. Printers generally find it easier to measure and control blacks and grays than mixtures of cyan, yellow, and magenta. By replacing CYM components with a single gray component, you can make it easier to control the density of the image. However, there are times when undercolor removal is not advised. It can be difficult to change the balance of CYM primaries after undercolor removal is performed. If a client decides at the last minute to change the color balance, it could lead to costly delays. Also, light gray screens composed of the CYM color components often look smoother than screens composed entirely of black dots.

We have seen that producing color from the desktop is a complicated matter. Unfortunately, a single chapter in a book is not going to give you a complete education in color production. It is important that you refer to the documentation that comes with desktop color software and hardware to learn what special capabilities the product offers. You should also work closely with your service bureau and print shop, especially if you are just starting out.

We'll explore those topics in the next two chapters.

# CHAPTER 8

# Working With a Service Bureau

If you produce your desktop-published pages on a 300-dpi laser printer, chances are that you own the output device. But desktop publishing users who own Linotronic imagesetters are rare. Though increasing numbers of companies have purchased the relatively inexpensive Linotronic 200SQ or 230, most Linotronic imagesetters are outside the price range of the typical desktop publisher. However, thanks to a new kind of business known as a desktop publishing service bureau, users can get all the advantages of high-resolution output without breaking the bank.

A "service bureau," in generic terms, is a company that offers use of computer-related equipment for people who cannot afford that equipment themselves. Slide-making service bureaus, for example, produce 35mm slides and transparencies for users who cannot afford their own film recorders. Video production service bureaus offer equipment for converting computer files into video animations. Desktop publishing service bureaus provide output on laser printers or high-resolution imagesetters, such as the Linotronic 300. They are also known as "PostScript" service bureaus for obvious reasons. The overwhelming majority of service bureaus that provide high-resolution output use Linotronic imagesetters.

Desktop publishing service bureaus are a relatively recent phenomenon. Linotype released its first PostScript imagesetter, the Linotronic 100, in 1986. Shortly after it was introduced, many companies saw an attractive business opportunity in offering imagesetter output to desktop publishing users. Some of these companies had their origins in the typesetting business, and saw the Linotronic imagesetter as a natural extension of their services. Others were involved in offset printing and photocopying, and again saw the imagesetter as a complement to their existing business. Some began as desktop publishing services, producing pages on 300-dpi laser printers before the imagesetter made high-resolution output possible. And other pioneering souls started from scratch, opening their doors as high-resolution output houses.

In addition to offering high-resolution output, many service bureaus provide related desktop publishing services, such as graphic design, typesetting, scanning, file conversion, training, photocopying, offset reproduction, or slide production. Service bureaus also vary widely in size and expertise. Some are small "mom-and-pop" shops run from modest storefronts, while others are massive operations with dozens of employees. The company's size, however, does not necessarily indicate the quality of its service or expertise. Many small service bureaus will go out of their way to help their customers get the best possible output.

In this chapter, we will offer guidelines on how to select and use a desktop publishing service bureau. Much of the advice in previous chapters will hopefully prove valuable in getting quality high-resolution output, whether the imagesetter is owned by you or someone else. But certain considerations are especially applicable when it comes to using a service bureau's Linotronic imagesetter.

## Choosing a Service Bureau

Using a service bureau without breaking your back or bankroll depends on a partnership. The service bureau should provide quality work and expertise, but the user needs to provide proper preparation to maximize quality and reduce delays. Knowing how the process works can help you go to a service bureau with the best of the pros.

The first step in selecting a service bureau is to talk with a customer service representative. In small service bureaus, this could be the same person who operates the imagesetter. Larger service bureaus have personnel specifically charged with communicating with the client. Either way, it is important that the customer representative be knowledgeable about the many subtleties involved in PostScript imagesetting. Many service bureaus are staffed by experienced personnel who will outline what to do and what not to do to get good quality output. Others, however, will print your files incorrectly and charge you for their mistakes.

During your conversation with the service bureau, try to determine how much they know about your hardware, software, and the types of documents you want to produce. Many service bureaus, for example, concentrate on producing output from Macintosh files, while others are more oriented toward IBM compatibles. Some encourage you to include scanned halftones in your jobs, while others prefer that you produce halftones by traditional means—a sign that they may be using older equipment that is too slow for large digital images. A good working knowledge of the software you use for output is also important. Such knowledge can be critical if problems develop.

Another important consideration is how the service bureau communicates with you. Are they friendly and courteous, or do they treat you with indifference? Do they use a lot of technical jargon that leaves you feeling confused, or do they answer your questions in a way that is easy to understand? When things go wrong, as they always do in service bureaus, the personnel need to tell you what went wrong and how to avoid it in the future. If they are knowledgeable, but poor communicators, you will never learn how to avoid your mistakes.

These key questions can help you identify service bureaus with experienced personnel:

- Ask what font libraries they support. Adobe and Linotype are the industry standard, but if they support more than one, it suggests they are dedicated to the field. Ask for a list of their fonts.

- Ask how they resolve font ID conflicts (see Chapter Four). A good service bureau should accept your screen font folders and know how to load your screen fonts in a way that prevents ID conflicts. If they can give you their screen fonts, they are aware of the problem and have devised a workable strategy.

- Ask what programs and versions they own. Do they own the program and version you want to work with? Ask what version of the LaserWriter and LaserPrep they would like you to use.

- Ask which version of the RIP they are using. Different versions of the RIP process at different speeds. If they charge extra for halftone-laden pages that take a long time to print, a faster RIP will be most cost-effective.

- Ask what resolution (dpi) they suggest you print. Typically, 1270 dpi is sufficient for text and most line art, but 2540-dpi resolutionor more  is preferred for most halftones and color separations. If they tell you that everything, including halftones, should be printed at 1270 dpi, then they are betraying a serious lack of expertise. If they tell you that it depends on what you are producing, you can be more confident that they know what they are talking about.

- Ask if they scan all disks for computer viruses. Service bureaus are among the most likely places where you can "catch" a virus. Find out what kinds of anti-viral tools they use.

- Ask them what screen frequencies to use for halftones. If one place tells you that everything should be printed at 90 lpi, they don't understand the spreading of inks that occurs on different papers in the printing process. If another place tells you it depends on what paper you are printing on—say 60 to 90 lpi for newsprint and 120 to 150 lpi for heavier coated stock—then you know the second place knows more then the first.

- Ask if they have an established procedure if problems arise. For example, if they can't print a file, will they call and ask you if they should continue, and estimate the charge to overcome the problem?

- Ask if they have a deep-bath or tabletop processor. Service bureaus equipped only with a table-top processor are ill-equipped to produce halftones and color separations. Also find out if they have a separate processor for film. Film requires higher developer concentrations than paper.

Once you've determined the service bureau's level of knowledge, you need to ask about other important considerations. These include:

**Price.** How much do they charge for the kind of work you want done? Charges typically vary depending on the number and size of pages in the job, output resolution, desired turnaround time, and whether the output is produced on resin-coated paper or film. Some service bureaus, especially those with older, slower RIPs, charge extra for jobs that take a long time to print, such as those that include halftones. If this is so, you are probably better off looking for a service bureau that has faster equipment. Also remember that experienced service bureaus often charge more because they spend time and money in testing and working out solutions to problems their clients encounter. A cheap price could end up costing you more in the long run.

**Payment terms.** Do they require that you pay for all work up front, or will they let you set up an account? You may need to fill out a credit application if you want them to bill you. Is there a minimum charge? Some service bureaus charge you for multiple pages as a minimum.

**Turnaround time.** How long does it take to produce a job? This can vary widely. Some service bureaus guarantee one- or two-day turnaround, while others produce jobs within two hours or less. Almost all service bureaus will do rush jobs on a while-you-wait or other short-term basis, but you could find yourself paying as much as 50 percent extra—or more—for the privilege. When asking about turnaround time, be sure they are referring to the type of output media you want. Some service bureaus, for example, reserve certain days of the week for film output. If you want film output on other days, you may have to pay a stiff premium.

**Output width.** Pages produced on the Linotronic 300 and 330 have a maximum width of 11.7 inches, while the 500 and 530 offer a maximum width of 18 inches. If you think you need the wider format, you should select a service bureau with one of the latter models.

**Color.** Are they set up to produce color separations? Can they handle process separations, or are they limited to spot color? Do they have a color scanner or printer? What kind of color proofing do they offer? This is an area that requires special expertise, so be sure to ask about color work they have done in the past.

**References.** Ask about the kinds of clients they serve, and try to get samples of their previous work, especially jobs that are similar to yours.

**File transport.** Do they have an electronic bulletin board so you can transmit files over a modem? Do they have a removable hard disk drive? If so, is it compatible with your removable hard drive (see below)? If you plan to bring diskettes in person, is the service bureau located within a reasonable distance, or will you need an overnight courier?

**Shipping.** Do they offer a delivery service, or do you have to rely on a courier (or yourself) to get your jobs back? Do you have to pay for delivery?

**Other services.** What kinds of ancillary services do they offer? These can include graphic design, typesetting, proofreading, scanning, file conversion, training, photocopying, offset reproduction, or slide production. Find out if they perform these services in-house or if they contract them out to other service bureaus.

# Preparing Your Files

Once you've selected a service bureau, the next step is to prepare your files for output. Here again, communication is important. To avoid problems that could delay your publication and add to the cost, always confer with someone at your service bureau before you begin.

Most service bureaus report that the biggest problems they encounter when producing output jobs involve fonts, especially fonts that fall outside the "LaserWriter 35" mentioned in Chapter Four. Here is where communication with your service bureau can be especially important. Some service bureaus, in fact, request that their customers use the bureau's fonts. Exchanging screen fonts with your service bureau is not a copyright violation, but this is not true of printer fonts. However, some service bureaus are authorized to sell packages of printer fonts from Adobe or other vendors, which ensures that you are both using compatible versions.

Sometimes, a customer may produce files with fonts that the service bureau does not have. One type family that causes particular problems in this area is Helvetica Narrow, which is found almost exclusively in laser printers. Instead of using Helvetica Narrow, Linotronic imagesetters use a similar typeface called Helvetica Condensed. If you try to print a file containing Helvetica Narrow on an imagesetter, the page will be printed in a poorly spaced version of Courier (Figure 8-1).

Even if you have what appears to be the same font as your service bureau, it could be from a different vendor. For example, if you compose your page using Bitstream's version of Futura and your service bureau prints the pages using Linotype's Futura, line and page breaks would change.

Another common problem is the font-ID conflict in Macintosh systems described in Chapter Four. One of the best ways to overcome this is to use a utility like Suitcase II or

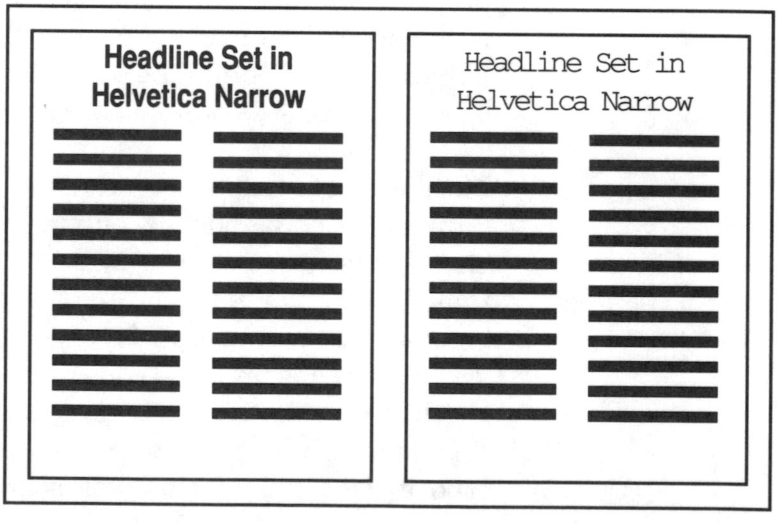

**Figure 8-1.** If you try to print a document with a font that has not been downloaded, the imagesetter substitutes Courier.

Master Juggler to install only the fonts you need for each job. Both programs are designed to help you insert and remove fonts without using the Font/DA mover and avoid the problems associated with font ID conflicts.

One way to verify that your fonts are compatible with your service bureau's equipment is to run a test file (Figure 8-2). Include a good representation of the different fonts you use. In addition, draw boxes, fill some in with different percentages of tints and draw lines of different widths. This will show if you have a font incompatibility problem. It will also show how the drawing elements and tints are different when printed from the imagesetter.

Print a draft copy on your own printer and send the file to a service bureau for output. Compare your printed test file with the service bureau's output. Some things, like tint densities, are guaranteed to change but line and page breaks must remain the same. If a problem occurs, try

**Figure 8-2.** By producing a test file such as this, you can be sure that a service bureau's imagesetter will be able to handle your output jobs.

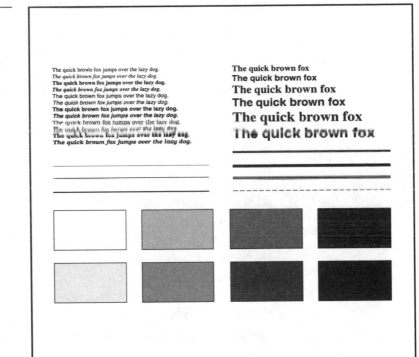

another test page, but this time take your screen font suitcase to your service bureau and request that they load them on their machine with Suitcase II or Juggler. In this way, your service bureau's system can access any unique ID numbers that have been assigned to your fonts.

Another important consideration is whether to produce PostScript output files or submit files in the native format of the software you are using. Submitting native files is generally as simple as copying them to disk—unless they are too big to fit. To create a PostScript file on disk, refer to the documentation for your software.

Some service bureaus will ask you to submit files in both PostScript and native format, along with constituent text and graphics files and any font files used in the job. You may be asked to use a file compression utility like Stuff-It on the Macintosh or Arc on the PC to store the files. This has two advantages: it reduces file sizes and groups the files together, ensuring that they won't get lost.

One program where it is essential to include constituent files is Ventura Publisher, which relies on them to store text and graphics. When copying Ventura Publisher files to disk, be sure to use the "Copy All" feature within the program instead of the DOS Copy command. Otherwise, the chapter will not open properly when the service bureau copies it from the diskette.

Whether you submit files in PostScript or native format, be sure to specify your service bureau's imagesetter in the "Print" dialog box, even if you don't actually print the file (Figure 8-3). You may need to give the Print command, then cancel it immediately, to save the appropriate setting in the dialog box. Also be sure to specify crop marks if you need them. You may have to choose a larger page size so the crop marks have room to print outside the text area. For example, on the Macintosh, "LetterExtra" will show crop

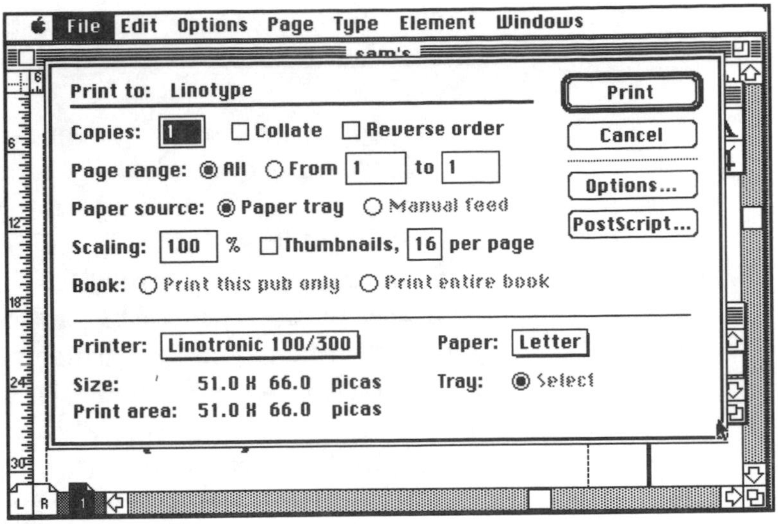

**Figure 8-3.** Before
submitting files for
output, be sure that you
have selected the
Linotronic imagesetter
as your output device.

```
 File  Edit  Options  Page  Type  Element  Windows
                          sam's

  Print to:  Linotype                          Print

  Copies:   1     Collate   Reverse order       Cancel

  Page range:  All   From  1   to  1
                                               Options...
  Paper source:  Paper tray   Manual feed
                                               PostScript...
  Scaling:  100  %   Thumbnails,  16  per page

  Book:  Print this pub only   Print entire book


  Printer:  Linotronic 100/300      Paper:  Letter

  Size:      51.0 H 66.0  picas     Tray:   Select
  Print area: 51.0 H 66.0  picas
```

marks on a letter-sized page just as "LegalExtra" and
"TabloidExtra" will allow crops to show on those pages.

If you are printing from the Macintosh version of Page-
Maker, your service bureau may request that you use the
Aldus Printer Description (APD) file when creating
PostScript output. This file allows the program to take
advantage of special features in the Linotronic imagesetter. It is supplied with the program, but the service bureau
may want you to use theirs. One reason why this is helpful
is that the Linotronic 300, with its maximum printing
width of 11.7 inches, cannot produce 11 x 17 inch tabloid
pages with normal-sized crop marks. However, a service
bureau can easily modify the APD so tabloid pages will
print with smaller crop marks.

One advantage with submitting PostScript files is that
they contain the information needed to specify the correct
font, reducing the possibility of font-ID problems. But if
your service bureau has installed the fonts specified in the
PostScript file, you can reduce the size of the file by
preventing the fonts from being downloaded. In Ventura,
change the fonts from "download" to "resident" in the Font

Utility dialog box. On the Macintosh, you simply hide your printer fonts in a separate folder.

PostScript files have their advantages, but they can also cause problems beyond the large amount of space required to store them. This is especially true if you submit PostScript files only. PostScript files are self-contained and can be difficult and expensive to modify. That means if you made a PostScript file to print a 13-page document it could cost you a lot of money to just print pages 5 to 7 unless you specify that in the print dialog box. In some cases, your service bureau may not be able to correct or modify a PostScript file. For example, if you want crop marks, but didn't choose the crop marks option, the service bureau can't add them for you. When problems occur, your service bureau will print out your page, charge you for it, and you will have to create another corrected PostScript file. So the PostScript file strategy doesn't work when you need the service bureau to correct the output the same day.

# Transporting Your Files

Once your files are prepared, you need to transport them to your service bureau. There are three ways to do this: copying them to a diskette, copying them to a removable hard disk or other mass-storage medium, or transmitting them by modem. The method you choose depends on your service bureau, the size of the files, and how soon you need to have them processed.

The simplest way to transport files is to copy them to diskette. The problem with this is limited storage capacity. If a file is too big to store on a single diskette, you can compress it using file compression software or divide it among multiple diskettes using a back-up program. This can be inconvenient, however, with especially large files. On older Macintosh computers with 3.5-inch diskettes, the maximum capacity is 800K. On the newer Macintosh SE/

30, IIx, IIcx, IIci, and IIfx, maximum capacity is 1.4 MB. Diskettes used in IBM-compatible computers come in both 5.25 and 3.5 inch varieties. The 5.25 inch diskettes range from 360K to 1.2 megabytes, while 3.5-inch diskettes range from 720K to 1.44 megabytes.

Most service bureaus have a 24-hour electronic bulletin board system that allows you to transmit files by modem. Modem transmission theoretically allows you to send files of any size, though there are practical constraints given transmission times and telephone costs. If you send a one megabyte file at 2400 baud, transmission time can run over an hour, and some service bureaus charge for long transmissions. Again, you can use a file compression program to reduce file sizes, but this may not be enough. Another solution is to use a high-speed 9600-baud modem. The problem here is a lack of a dominant standard for high-speed transmission, so you need to purchase a modem that is compatible with your service bureau's model.

The third alternative is to use high-capacity storage media, such as removable hard disks, portable hard disks, or erasable optical drives. Each type of media has advantages and disadvantages.

Removable hard disks can hold anywhere from 20 to 44 megabytes of data, but they suffer from a lack of standards. The drives generally fall into one of two incompatible categories: Bernoulli and SyQuest.

The Bernoulli technology, developed by a company called Iomega, was one of the first to support removable media on IBM-compatible computers, and has proven reliability in that environment. Early Bernoulli drives on the Macintosh had reliability problems, but these have been corrected in more recent products. Independent laboratory tests have indicated that Bernoulli drives tend to be more reliable than SyQuest drives. The drives are made exclusively by Iomega and work only with Bernoulli disks (Figure 8-4).

**Figure 8-4.** The Bernoulli Dual Drive from Iomega Corp. stores 44 megabytes on each disk.

SyQuest drives are produced by a variety of manufacturers that license the standard unit from SyQuest Technology (Figure 8-5). The technology is especially popular in the Macintosh environment, where it has become something of a standard. In addition to licensing the format to other companies, SyQuest Technology sells its own version of the drives with capacities up to 44.5 megabytes. Typically, the only difference among SyQuest drives is their INIT file, which is stored in the Macintosh system folder. If you need to send a SyQuest disk to your service bureau, you should also send a diskette containing the INIT.

Portable hard drives are similar to removable drives, except instead of transporting the disk, you transport the entire drive. They are designed to be moved and reconnected on a regular basis without loss of data. Most connect to the computer through the SCSI interface, making them especially suitable in the Macintosh environment. Their major advantage is compatibility: most will work with any Macintosh that has a SCSI interface.

Optical media represent the newest form of storage technology. They can store vast amounts of data on relatively small disks, but technological limitations restrict their usefulness for many end users. CD-ROM is useful as a distribution medium for software or large databases, but you cannot store your own data on the disks. WORM drives are useful for backing up important data, but they are ill-

**Figure 8-5.** SyQuest-based removable hard-disk drives, such as this model from CMS, store up to 44.5 megabytes per cartridge.

suited as transport media because you cannot erase the data once it has been recorded. However, a third kind of optical storage medium known as the erasable optical drive does present a viable alternative for users who need to transport their data.

Erasable optical drives use a combination of magnetic and optical technologies to store data that can later be erased or rewritten. Their high capacities and durability makes them ideal for transport purposes. Another advantage is compatibility; industry standards established for the drives ensure that a service bureau's drive will work with most of their customers' disks. However, they have a relatively short life expectancy of 10 years, limiting their usefulness as backup media. They are also quite expensive, with prices ranging from $5000 to $10,000 for the drives and $300 to $500 for the disks. Capacities range from 600 to 1000 megabytes.

Some manufacturers offer erasable drives with components manufactured by Ricoh that do not conform to industry standards. Their advantage is price: as little as $3000 for the drive and $230 for the disk. However, you cannot be sure that your service bureau will have a drive that can read the disks.

# Submitting a Job Ticket

When you submit a job to your service bureau, you need to include specifications for how the job should be produced. In some cases, specifications are recorded in the form of a job ticket, either on paper or in the form of an ASCII text file (Figure 8-6). Some electronic bulletin boards used in service bureaus allow you to enter job information in a menu-driven ordering system over the modem. In either case, this information should include:

- The software and version number you used to compose the work.

- The names of fonts used, and manufacturer.

- The names of the files you want printed and the numbers of the pages you want printed in each file.

- Whether it is a native application file (e.g. Ventura Publisher Chapter) or a PostScript file (using the print-to-disk function).

- Output size and whether crop marks are included.

- Whether registration marks are included.

- Whether you want film or paper output.

- Whether you want a positive or a negative.

- Whether you want the print on the emulsion side up or emulsion side down.

- Resolution stated in dpi (e.g. 635, 1270, 2540, 3386 dpi).

- Line screen stated in lpi (e.g. 90, 133).

**Figure 8-6.** A job ticket is used to list all the specifications for a particular output project.

**INVOICE**

**VG 2.0**
VISION GRAPHICS 2.0

Please Remit to
VISION GRAPHICS 2.0
2420 West Carson Suite 215
Torrance, CA 90501-3145
213.328.3339

Invoice Date_____

Bill To:_____

SERVICES REQUESTED

Company Name

Address

City          State          ZIP

Point of Contact

Phone Number

**OUTPUT**
☐ LINOTRONIC L300 - PAPER
☐ LINOTRONIC L300 - FILM
☐ QMS - PAPER
☐ QMS - TRANSPARENCY
☐ LASERWRITER

**INPUT**
☐ LAYOUT
☐ ILLUSTRATION
☐ WORD PROCESSING
☐ SCANNER - GRAPHIC
☐ SCANNER - OCR

**PICK UP**
☐ CUSTOMER DROP OFF
☐ VG PICK UP     MILAGE _____
☐ MODEM

**DELIVERY**
☐ CUSTOMER PICK UP
☐ VG DELIVER     MILAGE _____
☐ SHIP VIA _____

| TURNAROUND | TERMS | DUE DATE/TIME |
|---|---|---|
| ☐ NOW! +(100%)<br>☐ <4 HOURS +(50%)<br>☐ <8 HOURS +(25%)<br>☐ 24 HOURS | ☐ C.O.D.<br>☐ NET 10<br>☐ 2,10/NET 30 | |

SPECIAL INSTRUCTIONS/ ESTIMATE

| VG JOB NUMBER | CUSTOMER PO# | RESALE # |
|---|---|---|
| | | SR-AB |

**FONTS USED**

**APPLICATIONS**
☐ PageMaker
☐ Illustrator
☐ Quark Express
☐ FreeHand
☐ Letra Studio
☐ Design Studio
☐ Ready Set Go!
☐ Word
☐ OTHER _____

**Summary of Codes**

AS - Art Services     PL - Page Layout     SS - Scanning Services
LF - Linotronic-Film  QF - QMS-Film        WP - Word Processing
LP - Linotronic-Paper QP - QMS-Paper       VS - Vended Services
O - Other/Misc.       RS - Rush Services   ZP - LaserWriter

| UNIT CODE | NUMBER OF UNITS | UNIT PRICE | NET CHARGE |
|---|---|---|---|
| ① | | | |
| ② | | | |
| ③ | | | |
| ④ | | | |
| ⑤ | | | |

FILE NAME
① ② ③ ④ ⑤

| | |
|---|---|
| SUBTOTAL | |
| SALES TAX | |
| PICKUP/DELIVERY | |
| **TOTAL** | |

OK'D BY

RECEIVED BY

Please see additional terms and conditions on reverse

White- VG Work Order   Yellow- Invoice Copy
Pink- Return With Payment   Goldenrod- Customer Order Copy

- Color separation or composite.

- If they are color separated, how do you want them color separated? Should they be separated in the program that created the file or with a specialized utility like Adobe Separator.

- When you need the job (ask about rush charges).

- Will you pick-up or should they be shipped.

- The names of graphic files used in the job.

# Output Options

One key decision when producing output on a Linotronic imagesetter is the resolution to use. As we saw earlier, the Linotronic 300 is capable of producing output at 1270 or 2540 dpi. The Linotronic 330 offers a maximum resolution of 3386 dpi. Higher resolutions are preferred if you are producing color separations or scanned halftones. However, they are generally slower and more expensive. Lower resolutions are fine for text and simple line art.

Another consideration is whether to go to paper or film. As we discussed in previous chapters, film generally offers better output quality and can save a generation in the offset reproduction process. On the other hand, paper output is less expensive and makes it easier to proof the job. It is also allows you to perform further paste-up without going to the time and expense of stripping. Film output requires extra considerations, such as imagesetter calibration and processor settings, that are less crucial with paper output. For this reason, service bureaus that can correctly print to film are harder to find than those that print only paper.

## Producing Film Negatives

If you decide to print directly to film negatives you will have to find out how your printer wants the film. The terms used to describe this are confusing at first. "Emulsion," which is only on one side of the film, is the part that reacts to light and produces the image. "Right reading emulsion up" means that when you hold the film in front of you with the emulsion facing you ("Emulsion up") you can read the text "right." Conversely, "wrong reading, emulsion up" is when you hold the emulsion side toward you and the text looks like a mirror image or "wrong." The difference is important because exposing the film the wrong way can cause problems, such as shadows. Most printers want film "right reading emulsion down," but you should always ask. If you pick up your negatives from a service bureau check which side has the emulsion by scratching each side with a knife. The side that scratches is the emulsion side.

There are two methods you can use to produce film negatives from your desktop publishing or graphics program. The first method is simply to instruct your service bureau to output to film negatives, even though your document may call out positives. Your service bureau operator can always override the settings in your file by sending a "negative mode" command to the imagesetter.

The second method is to instruct your publishing or graphics program to "invert" your output. This is generally accomplished through the print dialog box (Figure 8-7). You can also control whether the emulsion is up or down, and right-reading or wrong-reading with the "Mirror" option. Table 8-1 lists the four possible settings for Invert and Mirror, and their effect on emulsion.

## Calibration Issues

Imagesetter calibration is an especially difficult task for many service bureaus. The difficulty involves choosing one of two conflicting calibration standards: film density and dot percentage. Density, or Dmax, is the "blackness" of the film when exposed at the maximum amount of black. This is the critical factor in creating film from which a printer can "burn" a printing plate. Typically, the desired DMax is

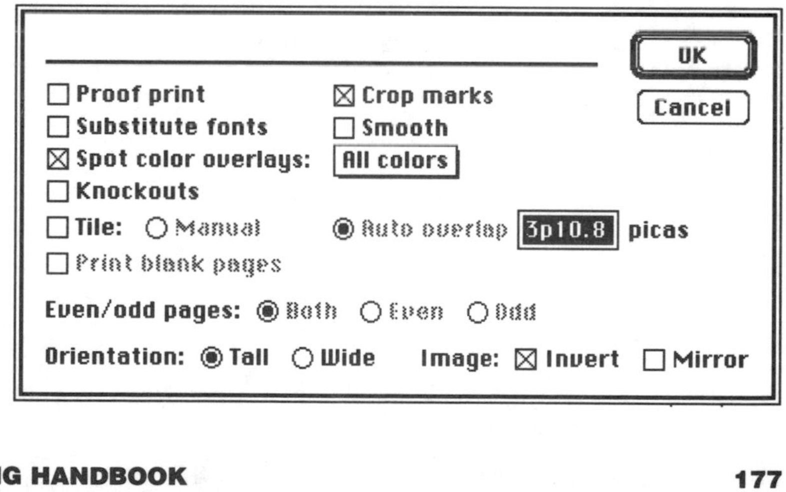

**Figure 8-7.** The Aldus print options dialog box allows you to produce negatives by checking the "Invert" option.

| Emulsion Option | Invert | Mirror |
|---|---|---|
| Right-reading, emulsion side up | On | Off |
| Right-reading, emulsion side down | On | On |
| Wrong-reading, emulsion side up | On | On |
| Wrong-reading, emulsion side down | On | Off |

3.5 on a densitometer, but 2.8 or above is acceptable. Ask your service bureau about the Dmax of their film negatives.

Another reason to find out about your service bureau's density settings is to ensure consistency in case you have to take a job to a different service bureau. This could happen if your service bureau's equipment breaks down, or if you have overflow work that needs to be done elsewhere.

With the conclusion of this chapter, we have followed the stages of production almost to completion. The last—and perhaps most important—stage is reproducing your Linotronic output on a printing press. We cover this process in the next chapter.

# CHAPTER 9

# Going to Press

A famous journalist named A.J. Leibling once said, "Freedom of the press belongs to those who own one." These days, he might well say that such freedom belongs to people who own a desktop publishing system. Computer-based publishing systems have given millions of users an inexpensive means of producing newspapers, books, magazines, and other forms of printed expression. Linotronic imagesetters provide a way to render desktop publishing output at high resolutions. But unless you have a way to reproduce that output, your insights and opinions won't get very far.

Fortunately, you don't have to purchase your own printing press to be a publisher. A look through your Yellow Pages will probably reveal dozens of companies that offer printing services in one form or another. These range from small quick-print shops to large commercial print houses with automated web presses.

Finding the right printer is an important part of any publishing job. Once you find that printer, you need to produce your camera-ready output in a way that brings the highest possible quality at the lowest possible cost. These considerations are especially important for users of Linotronic imagesetters. Whether it's a service bureau's imagesetter or your own, you have paid a premium for high quality output. All that effort and expense can be quickly negated if you ignore the reproduction process.

In this chapter we'll look at what it takes to go to press. We'll begin with the technology of page reproduction, following the process of offset lithography from beginning to end. We'll discuss the various kinds of inks and papers, hopefully providing insights into which ones to use for different kinds of printing jobs. Then we'll offer tips on how to produce camera-ready pages, including a discussion of how to produce impositions. Finally, we'll offer some suggestions for how to deal with your printer.

This last topic is an important one, because your printer can be a valuable asset in your quest to produce a good-looking publication. As we shall see, printing is a complex process, requiring detailed knowledge of inks, papers, and presses. This is especially true if you are interested in color. If you want the best possible printing job, the key is to talk with your printer before you begin a publishing project, not when it's nearly finished.

# Types of Reproduction

For most Linotronic users, reproduction means offset lithography. This process is widely used to produce magazines, books, newspapers, and other commercially printed material. It is also used in many quick-print shops for jobs where the quantity or quality requirements exceed the capabilities of photocopying machines. Other major printing technologies include letterpress and gravure, but these are not typically used to reproduce Linotronic output.

One form of reproduction to be avoided is the photocopier. These machines are great for reproducing laser printer output, but do not do justice to the high resolution of imagesetter output. Halftones in particular tend to suffer when reproduced on a photocopier. As we noted above, users of Linotronic imagesetters expect a certain level of quality in their work—a level of quality that cannot be matched by the photocopier.

**Figure 9-1.** The two types of offset press: sheetfed (below) and web (right).

Offset presses fall into two general categories, sheetfed and web (Figure 9-1). Sheetfed presses, true to their name, print on single sheets of paper sold in standard sizes, such as letter (8 1/2 x 11 inches), legal (8 1/2 x 14), or tabloid (11 x 17). Web presses print on large sheets of paper containing anywhere from four to 64 pages. Once the press run is complete, the sheets of paper are cut, folded, and stitched into bound pages.

Sheetfed presses, found largely in storefront quick-print shops, are used to produce newsletters, leaflets, flyers, and other publications that consist of single sheets up to tabloid size. They are suitable for small- to medium-sized print runs of 100 copies or more. Web presses, found in large commercial printing houses, are used for newspapers, magazines, books, and business forms.

# The Offset Lithography Process

Whether a job is produced on a sheetfed or web press, the process of offset lithography is pretty much the same. It begins with pages provided by the publisher. We call them "camera-ready" pages because they are photographed by a special kind of camera in the first step of the process. The camera produces large film negatives that contain a reversed image of the page.

## Stripping

At this point, the press operator engages in a laborious process known as stripping. Here, the negatives are taped into position on sheets of paper or plastic. If the job includes conventional halftones or other separate elements, the operator tapes them into the appropriate position (of course, if the page is produced on a Linotronic imagesetter, a digital halftone can already be in place). The pages themselves are arranged into impositions so they will be printed in the correct sequence. Impositions are relatively simple on a sheetfed press. On a web press, they can include as many as 64 pages. We'll discuss impositions more extensively later in this chapter.

## Bluelines

Once the negatives have been stripped into place, the press operator often creates a proof known as a blueline or brownline. This is done by placing the negatives over a photosensitive paper and exposing it to light. Wherever the light hits through the negative, the paper turns dark brown or blue. After the paper is exposed, it is cut and bound into a monochrome representation of the print job.

A blueline or brownline is your last chance to check a job before it goes to press. The proof serves mostly as a way to verify the positioning of pages, advertisements, halftones, and other elements. If you catch a mistake made by the printer, it is usually fixed free of charge. But if you see something you should have caught earlier—a misspelled word, a punctuation error, or the like—it will cost you extra to have the press operator strip in the corrected copy.

## Printing Plates

The next step is producing a printing plate. This is generally a thin sheet of metal or plastic with a light-sensitive chemical coating. The press operator places the negatives and a blank plate in a platemaking machine. Here, the plate is exposed to light, just as the proof was exposed earlier. But instead of causing marks to appear on the plate, the light hardens the chemical coating in the exposed areas. The operator then spreads a solution on the plate that removes areas not exposed to the light. The plate appears as a kind of bas-relief, with the exposed areas raised.

The plate is placed on a drum in the printing press. As the drum rotates, the plate picks up ink and transfers it to a second rotating drum covered with a rubber blanket. Ink is then transferred from the rubber blanket to the sheets running through the press. In simple monochrome printing, this happens once. In color printing, however, the sheets are run through the press once for each color to be applied. If necessary, the sheets are dried, turned over, and printed a second time on the reverse side.

## Finishing Operations

Once the printing is completed, finishing, and binding operations may be required. On a sheetfed press, this might include folding, cutting, collating, or stapling the pages. On a web press, finishing and binding operations are generally more extensive. Because the sheets contain multiple-page impositions, they must be cut and folded into signatures, and ultimately bound.

Your printer can tell you about the various options for binding and finishing your printed material. Common folds include four-page (one sheet folded in half), standard six-page (three sections, with the two ends folded over the middle panel), or six-page accordion. Common bindings include saddle-stitched, in which staples are placed in the spine of the publication, and side stitched, in which staples are placed on the edge near the spine.

# Ink and Paper

Any good recipe requires the correct mix of ingredients. In printing, the primary ingredients are ink and paper. Both come in many varieties, and both will have much to do with the final appearance of your print job.

Many variables influence the selection of both ink and paper. Important factors in the choice of ink includes the type of printing press, the paper to be used, and the ultimate use of the printed material. Another important factor is color. Inks used for spot color can be opaque or transparent, but inks used in process color must be transparent. In most cases, your printer will know what kind of ink to use for a particular project.

## Inks

If you are concerned about the environment, your printer may be able to offer environmentally safe inks made from non-toxic chemicals. Some inks are designed to dry easily with the help of ultraviolet or electron beam radiation. This reduces the emission of potentially harmful solvents into the air.

Numerous ink-related problems can arise in the printing process. These include distorted colors, mottling, and an effect called ghosting in which a reversed image of the page appears on another page. In most cases, a good press operator can correct these problems. This is a strong argument for carefully inspecting the printer's previous work before you hire them, especially work that is similar to yours.

## Paper

Papers used for offset reproduction also come in many varieties. At the bottom of the scale is newsprint, typically used in the newspaper industry because of its low cost. Heavier uncoated stocks are often used for books, newsletters, and other publications where you're looking for a compromise between expense and quality. At the top of the scale are chemically treated coated stocks, which typically offer the most faithful reproduction of the original document.

Many variables come into play when selecting a paper, including your budget, the kind of image you want to present, the presence or absence of halftones, the type of publication, and the type of output device used to create the original. One of the most significant choices involves coated vs. uncoated paper stock. Coated papers, while generally more expensive than uncoated, are a good choice for reproducing Linotronic imagesetter output. This is because the smooth texture of the paper faithfully reproduces the high-resolution features of Linotronic output.

### *Weight*

Another variable is the weight of the paper. Printers use two types of measuring systems, m-weight and basis weight. M-weight represents the weight of 1000 sheets at a particular size. If 1000 23 by 35-inch sheets of a book stock weigh 136 pounds, that paper is said to be a 136-pound stock. The more commonly used basis weight is the weight of 500 sheets at a certain size. Bond paper, for example, uses 17 by 22-inch sheets as the basis, while book paper uses a basis of 25 by 38 inches. A 20-pound bond stock is thus equivalent to a 50-pound book stock because the book paper uses a larger sheet size as its basis.

The weight of the stock influences a property known as opacity. Text printed on low-opacity paper tends to show through the other side, especially when held up to light. High-opacity papers are thus recommended for jobs where sheets are printed on both sides.

### Uncoated Stock

Uncoated papers fall into two categories, textured and smooth. Textured finishes come in several varieties, including linen, laid, and vellum. Linen finishes appear to have a finely detailed grid-like texture, and offer a professional look for text-intensive documents like newsletters and resumes. A laid finish appears to have many tiny ripples, and is also a good choice for newsletter publishers. Vellum finishes have finer ripples, and are popular for book production. Vellum offers a high degree of opacity and can make a book appear thicker than it would if printed with a smooth stock of the same weight.

### Coated Stock

Coated papers also come in several varieties. Standard glossy stocks offer a high degree of ink holdout, but can be hard to read because of light reflected from the page. The most expensive stocks use a dull coating, which provides excellent ink holdout without reflecting too much light. As a compromise, some publishers choose lightly coated matte stocks. These are easier to read than standard glossy coatings, but are much less expensive than dull coatings. On the downside, they offer less ink holdout than other coated papers. Matte coatings are popular among magazine publishers, especially those with large print runs. Dull coatings are popular for annual reports and other publications that require top-notch quality.

By talking with your printer, you can get a good idea of how a particular combination of ink and paper will handle your job. For example, if you anticipate a high degree of dot gain on a specific grade of uncoated paper, you can try to reduce the size of dots in any bit-mapped images on the page. You might also want to step down to a lower resolution for text, since some of the fine detail of the high resolution will be lost.

# Preparing Camera-Ready Copy

Now that we know more about the offset lithography process, we can produce camera-ready pages in a way that maximizes quality and reduces cost. One key consideration for users of Linotronic imagesetters is whether to produce pages on resin-coated paper or film. Both have their advantages and disadvantages.

The advantages of paper are cost and convenience. Most service bureaus charge less for paper output than for film. With paper output, you can easily proofread your camera-ready copy. And in some cases, you may not have a choice: many printers will not accept camera-ready copy in any other form.

Service bureaus charge extra for film output, but it is often worth the expense. By producing your pages as film negatives, you can bypass the first step of offset lithography. At each stage in the process, from camera-ready copy to final plate, your pages lose a slight bit of clarity. Skipping that first step removes a generation in the page's path to print. Film also does a better job of reproducing images with fine detail. You can also save money.

If you plan to produce halftones with screen frequencies of 120 lines or more, you should almost always print to film. If your printer won't accept film, find one that does. On the other hand, if your pages consist entirely of text and simple graphics, you may be better off going to paper. Proofreading is generally more important in text-intensive publishing projects, and paper output can be proofed directly. You can also save considerable money on high-volume jobs by going to paper.

Once you've decided on your output media, you need to prepare your pages so they can be properly "put to bed," which is publishing parlance for the act of sending a job to the printer. Again, as with every other aspect of the reproduction process, it is important to talk with your printer to learn about any requirements or suggestions that can improve the quality of the print job.

## Crop and Registration Marks

It is usually a good idea—and often a requirement—to include crop marks on your Linotronic output (Figure 9-2). Crop marks are short, narrow lines that indicate the boundaries of each page. They appear just beyond the four corners of the page, set in pairs at right angles. Crop marks make it easier for the press operator to position your pages and minimize the possibility for errors. They are especially important when printing to odd paper sizes.

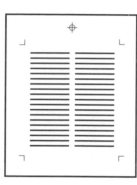

**Figure 9-2.** Crop marks (left and right) indicate the boundaries of the page. Registration marks (center) are used to align color separations.

Most desktop publishing programs include a printing option that automatically adds crop marks to the output. You can also create crop marks manually and position them on the page yourself. This is usually done with the line tool, a common function in many publishing and graphics packages. The key is to set the crop marks at 90-degree angles and keep them out of the active printing area. Some programs include a specialized line tool that produces straight horizontal or vertical lines. Others have a constraining function that restricts line drawing to horizontal or vertical. For example, in Ventura Publisher, you hold down the shift key while drawing the line to constrain it to the horizontal or vertical positions.

If you plan to produce color separations, you also need to include registration marks on your pages. These are small circles centered within slightly larger "plus signs." They are placed at the top and bottom of the page to assist the press operator in correctly aligning color separations. Most programs that produce color separations can automatically add registration marks.

## Bleeds

One special printing effect used in magazines and other media is the "bleed." In a bleed, a graphic element, usually a photograph or border, extends to the edge of the page (Figure 9-3). To create a bleed, you must provide camera-ready artwork in which the element that bleeds extends slightly beyond the page border. Bleeds usually involve extra expense because the printer must produce slightly larger pages, which are then trimmed.

**Figure 9-3.** In a bleed (right), an illustration extends to the edge of the page.

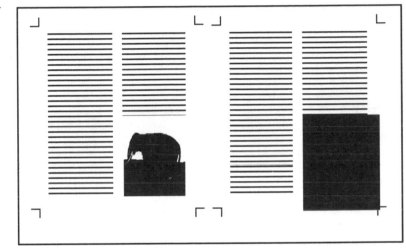

## Impositions

Creating impositions is a tricky process, but it can save a considerable amount of money on stripping services. Most people in the printing industry associate impositions with web presses. But you can also create simple impositions for sheetfed press work.

To get an idea of how impositions are created, try folding a sheet of letter-sized paper in half. Imagine it to be a four-page newspaper. Take a pen and label the pages: page 1 on the front, pages 2 and 3 on the inside, and page 4 on the back. Now unfold the sheet. On one side, you'll see pages 2 and 3 as you did before. But on the reverse side, pages 4 and 1 are next to each other, page 4 on the left and page 1 on the right (Figure 9-4).

What you have created is a four-page signature for a newsletter or newspaper. The reader (hopefully) begins with page 1 and continues through page 4. But to the press operator, page 4 comes first, followed by 1, 2, and 3. By adding more signatures, you can increase the number of pages as much as you want—in multiples of four. Many newsletters are created in this manner, except they are printed on tabloid-sized paper instead of letter size. The tabloid size corresponds to two letter-sized sheets, so each of the four pages in the signature measures a full 8 1/2 x 11.

**Figure 9-4.** Two common kinds of signatures used in commercial printing. In the eight-page signature (top), pages 1, 4, 5, and 8 are printed on one side of the sheet and the remaining pages on the other side. In the 12-page signature (bottom), all pages are printed on one side of the sheet. Sheets are then "tumbled" (turned over), inverted, and printed again so that the half with pages 1, 3, 6, 7, 10, and 12 is printed on the other side of the half with pages 2, 4, 5, 8, 9, and 11 (and vice-versa). Finally, the sheets are cut along the dotted line and folded.

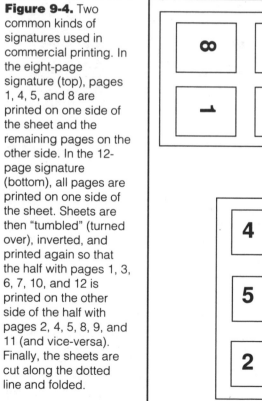
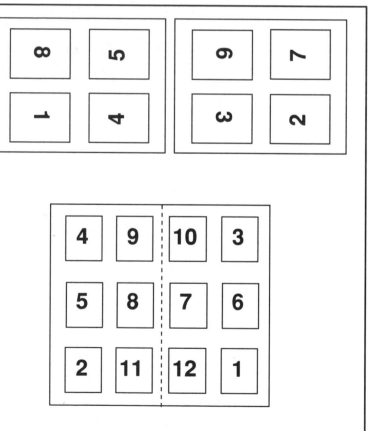

Because it fits within the 11 x 17 inch page size, this four-page imposition can be printed on a sheetfed press. But on a web press, impositions can include between four and 64 pages—in multiples of four. After the impositions are printed, they are folded and cut into signatures, which are then bound with other signatures into the book or magazine. Figure 9-4 shows some of the ways pages can be arranged in these larger impositions. But the general idea remains the same: fold a sheet of paper into the desired number of pages, then label them in sequence. When you unfold the sheet, the labels will show you the correct positioning of the pages.

As we noted in the section on the offset lithography process, impositions are usually created in the stripping process, which can get expensive. But with a little forethought, you can have a Linotronic imagesetter do at least part of the work of creating the impositions and save yourself some money in the stripping stage.

Of course, with a maximum page width of 11.7 inches on the Linotronic 300 and Linotronic 330, or 18 inches on the Linotronic 500 or 530, you will not be able to create complete 16-page signatures. But you can create smaller, self-contained impositions, or create sections of larger ones. The four-page signature we described above is relatively easy to create; you can also produce eight-page signatures in which each page is half-sized: 5 1/2 x 8 1/2. Even though these are true signatures, they can be produced on a sheetfed press, especially if the print shop has the necessary cutting, binding, and folding equipment.

The imagesetter can also produce sections of larger impositions. In Figure 9-4, we saw that an eight-page signature might have pages 6, 3, 4, and 5 on one side and pages 8, 1, 2, and 7 on the other. If you produce those two sets of pages separately on the imagesetter, the stripping job has been considerably simplified. Instead of stripping together eight pieces of film, the press operator need only deal with two.

When producing impositions on a Linotronic imagesetter, be aware of the way pages are printed. Pages normally come out lengthwise, as they do in a laser printer. But to economize on space—and service bureau costs—you want the pages turned sideways. Some desktop publishing packages, such as Aldus PageMaker, provide this kind of control.

One problem in creating impositions is arranging your pages in the correct sequence. Most desktop publishing programs were first designed when the laser printer was the only means of output. True to their origins, they still handle printing jobs as a series of pages in numerical sequence, starting with page 1. With many programs, you can rearrange pages and page numbers, but not without a great deal of difficulty.

A software product called Impostrip, from Ultimate Technographics in Montreal, Canada, can be a valuable tool for creating impositions (Figure 9-5). This product comes in several versions that work with specific desktop publishing packages, including Ventura Publisher, Aldus PageMaker, QuarkXPress, and the company's own publishing software. Its primary function is to convert PostScript files created by these programs into user-defined signatures. Unfortunately, it is quite expensive—$4000 for each of the program-specific versions, or $20,000 for a single all-encompassing version. However, some service bureaus have their own copy of the Impostrip software that they use to produce signatures from customers' jobs.

Mastering the art of imposition can save considerable amounts on printing costs while improving the appearance of your publication. Some commercial printers, for example, charge the same amount for producing a 12-page signature as they do for a 16-page signature. By effectively combining signatures, you can get more pages printed for the same amount of money.

**Figure 9-5.** Impostrip, from Ultimate Technographics, allows users of standard desktop publishing packages to produce impositions.

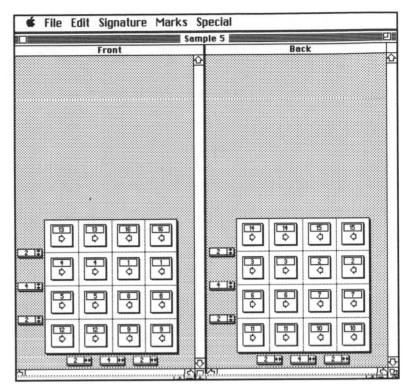

Impositions are also important in color publishing. Whether you are producing spot or process color, it costs extra every time you run a set of sheets through a printing press. But if you print an extra color on one page, you can print it on all other pages on the same side of the signature for no extra cost. If you charge extra for color ads in your magazine or newspaper, your advertiser can underwrite the cost of printing other pages in color.

# Working With Your Printer

Your printer can be a valuable asset in your quest for quality and economy. You should thus make your selection carefully. This means asking for references and looking at samples of the printer's previous work. Be sure to ask for samples that are similar to the type of work you want handled.

Like every other business, printers have their specialties. Some specialize in books, others in newspapers or magazines, still others in business forms. Smaller storefront print shops tend to be more generalized, handling resumes, flyers, newsletters, direct mail pieces, wedding invitations, and other items with relatively modest print runs.

### Selecting a Printer

The first step in finding a printer is to send out queries specifying the kind of work you want done and requesting a price estimate. The specification should include the number of copies, type of paper, any necessary finishing or binding, and other special instructions. The estimate should include costs for paper, printing, binding and finishing, and shipping or delivery. It should also indicate the printer's payment terms. Find out if you can save on stripping charges by submitting film negatives and/or camera-ready artwork arranged into impositions. You may be able to save on paper costs by providing your own.

You should also find out about the printer's turnaround time—the typical amount of time it takes to complete a job—and any other services that may be offered. Some print shops, for example, offer mailing services. Some even include their own Linotronic imagesetters.

Some of this information can be obtained over the phone, but any formal requests for bids—and estimates—should be put in writing.

### Specifying Jobs

When evaluating responses to your queries, be sure that the printer is responding to your specifications. Some may make minor modifications in paper stock or trim size to accommodate their equipment. If the changes are acceptable to you, fine. If not, there are probably plenty of printers who can meet your exact specifications. Whatever happens, be sure to get all of the printer's estimates and promises in writing.

Before you make your final decision, talk to the printer's customer service representative if you haven't done so already. This person will be your primary contact with the printer. Ask yourself if the service rep is someone you can work with: someone who's honest, personable, and knowledgeable about printing in general and your kind of printing job in particular.

As we have noted, a good printer representative can be a valuable asset. Don't hesitate to call on them if you have questions or need help in preparing your camera-ready artwork. They won't do your work for you, but they can offer advice that can save valuable time and money.

It is also important to treat your printer with professional respect. This means paying your bills on time and sticking to deadlines. Many printers schedule their work carefully to keep their presses—and employees—working at a steady rate. If you miss a deadline without giving adequate notice, the printer could face an unexpected slack period, followed by a period of frenzied activity as they try to fit your job in. That is not a prescription for quality work.

Producing quality means starting at the end—the offset printing press—and working your way back to the beginning. When you start your publishing project, keep in mind that your ultimate product is the printed page, and plan accordingly.

**CHAPTER 10**

# Special Tips and Techniques

In previous chapters, we've looked at several general topics of importance to users of Linotronic imagesetters. In this final chapter, we'd like to leave you with some practical tips and advice for getting the most out of a Linotronic imagesetter. Many of these tips tell you how to accomplish a task more efficiently or cost-effectively than you have previously. Others tell you how to perform a task you might never have thought of.

Chances are, once you've gained some experience outputting jobs on a Linotronic imagesetter, you'll be able to come up with some tips of your own. We encourage you to share that advice with other users by submitting them to a publication, attending a users group meeting, or letting your service bureau know. If you think of it, drop us a line at the address in the front of this book. We'd love to hear from you!

## Saving Money on Service Bureau Charges

Most desktop publishing service bureaus base their output charges on the amount of paper or film used in the job. It makes sense, because these consumables represent one of the greatest expenses in operating a service bureau. But with a few common-sense techniques, you can save yourself some money. Not a lot, maybe, but it can add up over time or on large typesetting jobs.

The general rule in imagesetting is to minimize the amount of empty space. You pay the same amount for each page, whether there's only a line of text or three full columns with images. Of course, you'll always need a certain amount of white space (or black space in the case of negatives) for crop marks, registration marks, and other symbols. But why pay for more than you need?

## Selecting Page Size and Orientation

The first issue to consider is the size of paper (or film) used in the service bureau's imagesetter. Usually, imagesetters like the Linotronic 300 use paper that's about 11.7 inches wide. In normal operation, pages come out head first, just as they come out of a laser printer. So if you're using a typical 8 1/2 by 11-inch document size, you should orient the output so that the tall (11-inch) dimension runs the width of the roll of paper in the imagesetter, rather than

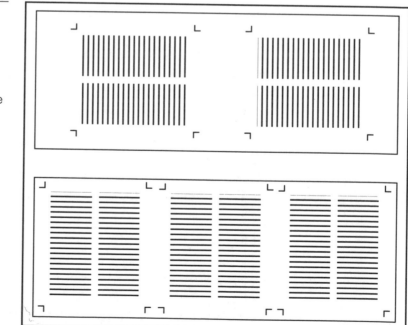

**Figure 10-1.** As a default, the Linotronic imagesetter produces pages head-first (top). But by producing pages sideways (bottom), you can save film.

lengthwise, which would otherwise be normal (Figure 10-1).

In Aldus PageMaker, you select the appropriate page orientation for imagesetter output by specifying "Letter Transverse" as the Paper type from the "Print to" dialog box (Figure 10-2). This paper size is available only if you have selected an output device such as the Linotronic 300 that permits this size. Do not confuse the Letter Transverse paper size with the landscape (or wide) orientation for your document. The orientation of your page—tall versus wide—is independent of the direction it is facing when it emerges from the imagesetter.

Another way to maximize use of paper is to print small pages two-up (side-by-side). This is useful for producing books and booklets that are 5 1/2 inches wide or less (Figure 10-3).

If you use PageMaker, here's a semi-manual way to produce two-up output for small page sizes. This method uses PageMaker's facing pages feature to output left/right page pairs. You must use manual tiling. Set the zero point on the ruler at the upper left corner of the left-hand page. Select

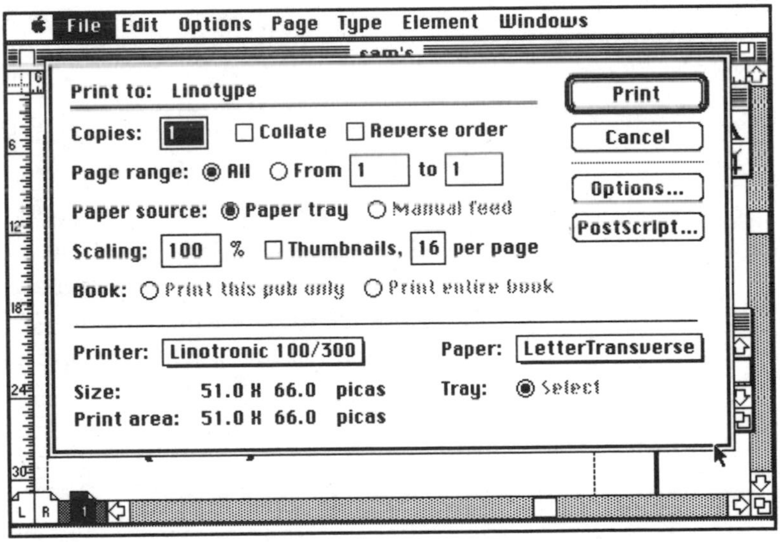

**Figure 10-2.**
PageMaker print dialog box with "Letter Transverse" selected.

**Figure 10-3.** By printing small pages two-up, you can save on service bureau charges.

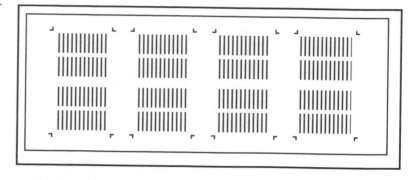

Letter Transverse paper size, then print the left-hand page individually. You'll find that the corresponding right-hand page comes out as well. Repeat this process for each left/right pair in your publication.

## Utility Programs

Another way of producing two-up output is to use Impostrip, a program from Ultimate Technographics in Montreal (see Chapter Eight). This software package positions page output side-by-side and creates impositions. Signatures created by the program can be used by your press operator to burn a printing plate without a lot of expensive stripping procedures. Versions are available for Ventura Publisher, PageMaker (PC and Macintosh), QuarkXPress, and Ultimate's own publishing software for about $4000 each. A service bureau version that works with all packages costs about $20,000.

If your service bureau has a Linotronic 500 or 530, it can produce output up to 18 inches wide. You can thus use the techniques outlined above for pages larger than 5 1/2 by 8 1/2 inches. One other point to keep in mind: if your PageMaker 4 document has blank pages in it, save yourself the expense of sending these to your service bureau. The Aldus package has a feature that lets you suppress blank pages automatically.

## Make Use of "Blank" Space

If you produce a magazine, newspaper, or other type of publication using a service bureau's imagesetter, you may be missing an opportunity to get some "free" typesetting jobs done. Specifically, you should always look for small jobs such as business cards or logos that can be inserted in otherwise empty space in your "main" job.

For example, if your publication contains advertising, it is likely that you are leaving blank spaces in your pages where ads can be stripped in by your print shop. Why not use those blank spaces for other jobs? It is a simple matter to drop in a business card, corporate logo or something even more substantial if space permits (Figure 10-4). Similarly, if you leave a space for color photos or other artwork that is handled conventionally, this is another opportunity for free typesetting.

There are two different approaches for making use of this free space. First, you can use your desktop publishing program to produce the "bonus" document in the same file as the main document. Alternatively, you can produce the free document in a separate file and then import it as an EPSF file into the main document file. This second ap-

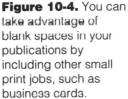

**Figure 10-4.** You can take advantage of blank spaces in your publications by including other small print jobs, such as business cards.

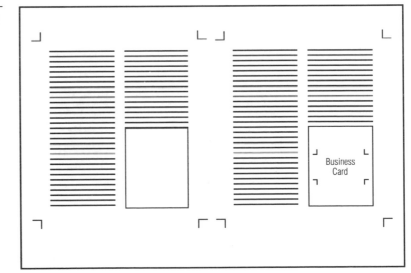

proach allows you to use different programs for the two different documents. It also allows you to produce automatic crop marks for the imported document.

To produce accurate crop marks, make sure you set up your document size to the exact physical dimensions. For example, if you are producing a business card, set up your page size as 3.5 inches wide by 2 inches high. You can do this using the "Custom" option in PageMaker's Page Setup menu. In Ventura Publisher, modify the size of the Underlying Page using the Sizing & Scaling option from the Frame menu.

When the job comes back from your service bureau, just clip out the inserted items before you have the pages printed. Be sure to leave a sufficient border, along with crop marks, around the inserted items.

# Modifying Aldus PageMaker APD Files

If you use Aldus PageMaker, you are probably aware that the program installs printer description files that tell it specific information about specific printers. The APD file for Linotronic imagesetters is called Linotronic 100/300.apd. If you know what you are doing, you can modify some of the parameters in this file to suit your taste.

Make a copy of the file, then open the copy in text-only mode with your word processing program. You will find a list of statements (Figure 10-5). Each statement begins with the at-sign (@) character, and is followed by the name of a parameter and its value in quotation marks. Most of the parameters in an APD file should not be changed, but there are a few values you may want to tinker with. For example, the parameter @DefaultScreenFreq modifies the default halftone screen for tints and shades produced with PageMaker's graphics tools. You may want to change this to 85, 100, 120, or 133, depending on which model

```
@Comment: "Linotronic 100/300.apd" for Linotronic 100 and 300 imagesetters.
@Comment: Aldus Printer Description (APD) file
@Comment:          $Revision: 3.7 $
@Comment:          $Date:  09 Jun 1988 15:57:46  $
@Comment: This APD produced for use with Aldus PageMaker 4.0 or earlier version.
@Comment: Keyword values are currently limited to 255 characters in length.
@FormatVersion: "1.0"
@Product: "(Linotype)"
@PSVersion: "(38.0)"
@PSRevision: "1"
@Comment: This resolution describes the Linotronic 100
@DefaultResolution: "1270 1270"
@Resolution: "1270 1270"
@AllowComments: "true"
@AllowVMQuery: "true"
@VariablePaperSize: "true"
@Comment:  "FreeVM" value is equivalent to vmstatus "maximum" minus "used"
values
@FreeVM: "170000"
@WorkingMem: "45000"
@AldusHeaderMem: "34000"
@Comment: Example of declaration of a downloadable font's memory requirement:
@Comment: @FontMem "Benjo-Light": "82600"
@Comment: PatchFile is sent right after %%EndComments, AldusPatchFile right
before %%EndSetup
@PatchFile: "userdict /AldusDict known {(A previous version PageMaker header is
loaded.) = flush} if"
@AldusPatchFile: ""
@ExitServer: "serverdict begin exitserver "
@Password: "0"
@DefaultScreenFreq: "90"
@DefaultScreenAngle: "45"
@DefaultScreenProc: "{abs exch abs 2 copy add 1 gt
{1 sub dup mul exch 1 sub dup mul add 1 sub} {dup mul exch dup mul add 1 exch
sub}ifelse}"
@End
@ScreenFreq: ""
@ScreenAngle: ""
@ScreenProc: ""
@InvertScreenProc: "{abs exch abs 2 copy add 1 gt {1 sub dup mul exch 1 sub dup
mul
add 1 exch sub}{dup mul exch dup mul add 1 sub} ifelse}"
@End
@Transfer: ""
@InvertTransfer: "{1 exch sub}"
@NormalizedTransfer: "{
mark
1.0 1.0 .92 .76 .64 .54 .44 .36 .28 .2 .1 .0
counttomark dup 3 add -1 roll exch
```

imagesetter you are using and what your printing requirements are. Other defaults you may want to change are @DefaultPageSize and @DefaultFont.

Also present within an APD file are the dimensions, in points, of each of the page sizes available to that particular output device. If you wish to add your own custom page sizes, for example, 7 x 9 inches for a book of this size, you can do this by giving a name to your new size, and adding statements in three different places in the file (ee Figure 10-6). The three statements we would add to create a new page size for this book would be:

@PageSize Book: "statusdict begin 504 648 0 0 setpageparams end"

@PageRegion Book: "0 0 504 648"

@PaperDimension Book: "504 648"

---

**Figure 10-6.** By including these statements in the APD file, you can add new page sizes.

```
@Comment: PageSize options appear in the "Paper" list box in the Printer-specific
dialog.
@DefaultPageSize:                        "Letter"
@PageSize Letter:                        "letter"
@PageSize LetterExtra:                   "statusdict begin 684 864 0 1
setpageparams end"
@PageSize A4:                            "statusdict begin 596 842 0 1
setpageparams end"
                                •
                                •
                                •
@Comment: PageRegion gives the printable area of each paper option.
@PageRegion Letter:                      "0 0 612 792"
@PageRegion LetterExtra:                 "0 0 684 864"
@PageRegion A4:                          "0 0 595.28 841.89"
                                •
                                •
                                •
@Comment: PaperDimension gives the total paper size of each option.
@PaperDimension Letter:                  "612 792"
@PaperDimension LetterExtra:             "684 864"
@PaperDimension A4:                      "595.28 841.89"
```

# Extracting Ventura Pages From PostScript Files

As we mentioned in Chapter Eight, it is often preferable to produce a PostScript file to send to a Linotronic service bureau, rather than sending the native document file. This method allows the service bureau to download the file directly without having to enter the application.

One problem with this situation is that it can be difficult to extract a single page or page range from a PostScript disk file that contains a multiple-page document. For example, you may want to reprint a particular page that is missing because the output device ran out of paper.

Here's a technique for extracting a page range from a Ventura Publisher PostScript file. Once you create the file, all you'll need is a word-processing program to make the technique work.

Because PostScript files are saved in straight ASCII text format, your word processor must be able to read and write ASCII files. If you try saving the file in your word processor's standard document format, it will not be capable of printing anything.

When Ventura produces a PostScript file via the Print to File command, it looks something like the one in Figure 10-7. The first section, known as a "preamble," is common to all PostScript output files. Following the preamble are the commands to produce the actual pages. You can locate the end of the preamble and the beginning of the first output page by searching for the first occurrence of the string "%Begin page." Each subsequent page will also begin with the "%Begin page" line. Thus, to extract a desired page or range of pages from the file, you just delete the pages you do not need.

Using your word processor's Find function, move to the first

**Figure 10-7.** To delete a page or pages from a Ventura Publisher output file, delete everything between the %Begin page and %End page commands.

```
%!PS-Adobe-1.0
%%Title: GEM Document
%%Creator: GEM
%%Pages: (atend)
%%BoundingBox: 0 0 611 791
%%EndComments
% Copyright (C) Digital Research, Inc. 1986-1988. All rights reserved.
systemdict /setpacking known
                                        •
                                        •
                                        •

%Begin page
UserSoP
greset -300 3599 2850 3599 2850 -301 -300 -301 np mto lto lto lto clip np
                                        •
                                        •
                                        •
greset -300 3599 2850 3599 2850 -301 -300 -301 np mto lto lto lto clip np
%End page
showpage svobj restore gr
gs /svobj save def
%Begin page
UserSoP
```

%Begin page line in the file. Here you can mark the beginning of the page to be deleted using your word processor's cut-and-paste functions. Then use another Find command to get to the next %Begin page operator.

If you are deleting just the one page, mark the end of the block immediately before the %Begin page command. If you want to delete several consecutive pages, keep using the Find command until you get to the end of the last page you want to remove. Once the block has been properly selected, you can delete it. You can then jump ahead to delete or restore subsequent pages.

In this manner, you should be able to isolate only the pages that you want produced. This will save you the trouble of having to load your Ventura document again to produce a new PostScript file.

In cases where you cannot reconstruct the Ventura document in its original format, this technique will provide a valuable insurance policy for your archived documents. And your service bureau will be able to extract pages from your file without having to produce the entire document, thus saving additional expenses for the service bureau or customer.

# Producing Masters for Three-Up Labels

If your office is anything like ours, you've learned the value of three-up adhesive labels that can be fed through a plain paper copier or laser printer. Unlike pin-fed pressure-sensitive labels used with dot-matrix printers, these labels come on 8.5 x 11 inch sheets, usually with 33 labels per sheet (11 columns by three rows).

Although many desktop publishing users feed these label sheets through their laser printer, we avoid this for several reasons. First, laser printers cannot print to the edges of the paper, and as a result, at least three and sometimes six labels on each page are wasted, depending on the image size and position. Second, label sheets can sometimes get jammed, or worse yet, labels can come off and adhere to the printer's rollers. Third, laser printer toner often does not adhere well to laser labels, and as a result it tends to rub off.

A much better solution is to produce the original on a Linotronic imagesetter and then reproduce it on label stock fed through a copier. Photocopiers can deposit toner on the entire page, and the toner seems to adhere better. And feeding label stock through a copier is usually safer than feeding it through a laser printer.

The advantage of producing the original on an imagesetter

is that it can cover the entire surface of an 8 1/2 x 11 inch page. The quality is also much better. This is especially useful for creating masters that are frequently reproduced, such as return address labels or company logos.

Many software packages can produce masters for three-up labels, but our favorite is Aldus Freehand. Using Freehand's text and graphic tools, create the first label in the upper-left corner of the page. Then group the label's elements and clone the image (Figure 10-8).

Using Freehand's precise Move Elements function, shift the cloned label exactly one inch below the first. Then use the Duplicate function nine more times to reproduce each label and automatically position it.

Next, group all 11 labels in the first column, clone the column, and move it exactly 17 picas to the right (one of the reasons we like the picas/points measurement system is that it lends itself to division by three). Repeat this process to produce the right-hand column of labels.

This process works with more than just one-inch address labels. Many office supply stores sell adhesive labels that come 4, 8, or 12 to the page rather than 33 per page. These larger formats are useful for diskette labels, catalog enve-

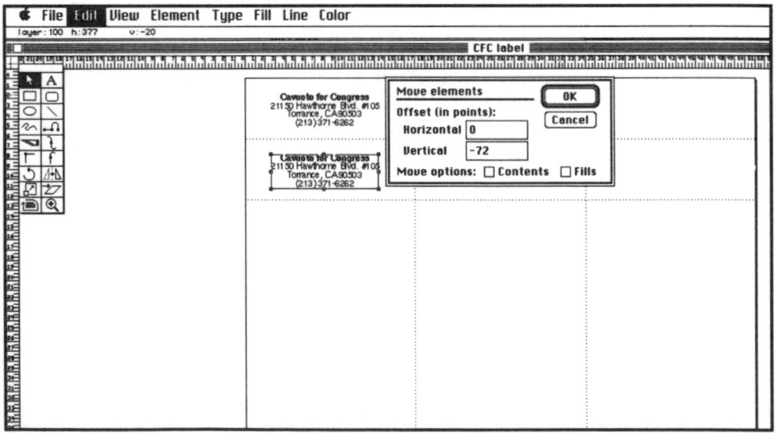

**Figure 10-8.** Three-up labels produced in Aldus Freehand for output on a Linotronic imagesetter.

**SPECIAL TIPS AND TECHNIQUES**

lope return addresses, and makeshift report covers. You can also buy clear acetate labels in a number of sizes.

# Creating 35-mm Slides on an Imagesetter

In this book, we have looked largely at desktop publishing applications of Linotronic imagesetters. Users of desktop presentation software, such as Aldus Persuasion or Microsoft PowerPoint, can also use a Linotronic imagesetter to produce slides for their presentations. There are numerous slidemaking service bureaus that produce 35mm color slides for clients using color film recorders from manufacturers such as Agfa Matrix and Mirus. While a Linotronic imagesetter cannot produce slides with the quality of a dedicated slidemaker, it may be convenient and cost-effective to produce certain slide presentations with it.

To use a Linotronic imagesetter for producing slides, you must specify film rather than paper output. You should also specify positive, as opposed to negative, output. And you should request the highest available output resolution; 2540 dpi on the Linotronic 300, or 3386 dpi on the Linotronic 330 imagesetter.

You must make sure that the images you send to the imagesetter are sized correctly to fit within the frame of a 35mm slide. One way to accomplish this is to specify an appropriate reduction factor from the Page Setup dialog box on the Macintosh. A better technique is to take advantage of your software's ability to produce miniature versions of slides. In this manner, you can group six or more slides on a single page of film—a considerable cost savings over slidemaking services.

You will have to manually cut each page into the individual slides and mount each slide in a frame. And, of course you are limited to one-color (black) output. But for low-cost, quick-turnaround presentations, a Linotronic imagesetter may be a viable alternative to film recorders.

These techniques conclude our advice on making the most effective use of a Linotronic imagesetter. Again we encourage you to experiment with your own ideas.

If we can leave you with one final thought, remember that the field of desktop imaging is advancing rapidly. We hope you will not treat this book as the last word on the subject, but rather, as a starting point for you to discover the power and quality of Linotronic imagesetters.

# APPENDIX

# Products and Vendors

## Graphics Software

**Adobe Illustrator**
Adobe Systems Inc.
1585 Charleston Rd.
Mountain View, CA 94039
(415) 961-4400

**Aldus Freehand**
Aldus Corp.
411 First St. South, Suite 200
Seattle, WA 98104
(206) 622-5500

**Arts and Letters**
Computer Support Corp.
15926 Midway Rd.
Dallas, TX 74244
(214) 661-8960

**Cliptures**
Dream Maker Software
7217 Foothill Blvd.
Tujunga, CA 91042
(818) 353-2297

**ColorSep/PC**
Ozette Technologies
800 Fifth Ave., #240
Seattle, WA 98104
(206) 747-3203

**ColorStudio**
Letraset USA
40 Eisenhower Dr.
Paramus, NJ 07653
(201) 845-6100

**Corel Draw**
Corel Systems Corp.
Corel Building, 1600 Carling
Ottawa, Canada, K1Z 7M4
(613) 728-8200

**DeskPaint/DeskDraw**
Zedcor
4500 E. Speedway #22
Tucson, AZ 85712
(602) 881-8101

**Digital Darkroom**
Silicon Beach Software
9770 Carroll Center Rd. #J
San Diego, CA 92126
(619) 695-6956

**Fontographer**
Altsys Corp.
720 Ave, F, Suite 108
Plano, TX 75074
(214) 424-4888

**FontStudio**
Letraset USA
40 Eisenhower Dr.
Paramus, NJ 07653
(201) 845-6100

**FullPaint**
Ashton-Tate
20101 Hamilton Ave.
Torrance, CA 90502
(213) 329-8000

**GEM Artline**
Digital Research Inc.
60 Garden Court
Monterey, CA 93942
(408) 646-6208

**Gray FX**
Xerox Imaging Systems
535 Oakmead Parkway
Sunnyvale, CA 94086
(408) 245-7900

**HotShot Graphics**
SymSoft Corp.
Call Box 5
924 Incline Way
Incline Village, NV 89540
(702) 832-4300

**ImageStudio**
Letraset USA
40 Eisenhower Dr.
Paramus, NJ 07653
(201) 845-6100

**LaserTalk**
Adobe Systems
1585 Charleston Rd.
Mountain View, CA 94039
(415) 961-4400

**MacDraw/MacPaint**
Claris Corp.
440 Clyde Ave.
Mountain View, CA 94043
(415) 987-7000

**Micrografx Designer**
Micrografx Inc.
1303 Arapaho
Richardson, TX 75081
(214) 234 1769

**PC Paintbrush**
ZSoft Corp.
450 Franklin Rd., #100
Marietta, GA 30067
(404) 428-0008

**PhotoMac**
Avalon Development
1000 Massachusetts Ave.
Cambridge, MA 02138
(617) 661-1405

**Photoshop**
Adobe Systems
1585 Charleston Rd.
Mountain View, CA 94039
(415) 961-4400

**Picture Publisher**
Astral Development Corp.
Londonderry Square, Suite 112
Londonderry, NH 03053
(603) 432-6800

**Publisher's Type Foundry**
ZSoft Corp.
450 Franklin Rd., Suite 100
Marietta, GA 30067
(404) 428-0008

**SmartArt**
**Adobe Systems**
1585 Charleston Rd.
Mountain View, CA 94039
(415) 961-4400

**Streamline**
**Adobe Systems**
1585 Charleston Rd.
Mountain View, CA 94039
(415) 961-4400

**SuperPaint**
**Silicon Beach Software**
9770 Carroll Center Rd. #J
San Diego, CA 92126
(619) 695-6956

## Desktop Publishing Software

**Aldus PageMaker**
**Aldus Corp.**
411 First St. South, Suite 200
Seattle, WA 98104
(206) 622-5500

**DesignStudio/Ready,Set,Go!**
**Letraset USA**
40 Eisenhower Dr.
Paramus, NJ 07653
(201) 845-6100

**FrameMaker**
**Frame Technology**
1010 Rincon Circle
San Jose, CA 95131
(408) 433-1928

**Interleaf Publisher**
**Interleaf Inc.**
10 Canal Park
Cambridge, MA 02141
(617) 577-9800

**PFS:First Publisher**
**Software Publishing Corp.**
1901 Landings Dr.
Mountain View, CA 94043
(415) 962-8910

**QuarkXPress**
**Quark Inc.**
300 S. Jackson St., Suite 100
Denver, CO 80209
(303) 934-2211

**Ventura Publisher**
**Ventura Software Inc.**
15175 Innovation Dr.
San Diego, CA 92128
(619) 673-7537

## Font Software

**Adobe Type Manager**
**Adobe Systems**
1585 Charleston Rd.
Mountain View, CA 94039
(415) 961-4400

**FontMaster/Font Wizard**
**Varityper, Inc.**
11 Mt. Pleasant Ave.
East Hanover, NJ 07936
(201) 887-8000

**Fontware**
**Bitstream Inc.**
Athenaeum House
215 1st St.
Cambridge, MA 02142
(617) 497-6222

**Linotype Type Library**
**Linotype Co.**
425 Oser Ave.
Hauppauge, NY 11788
(516) 434-2717

**Master Juggler**
**Alsoft Corp.**
P.O. Box 927
Spring, TX 77383
(713) 353-1510

**Treacyfaces**
**Treacyfaces**
111 Sibley Ave., 2nd floor
Ardmore, PA 19003
(215) 896-0860

## Other Software

**BarneyScan XP**
**BarneyScan Corp.**
1125 Atlantic Ave.
Alameda, CA 94501
(415) 521-3388

**GEM**
**Digital Research Inc.**
60 Garden Ct.
Monterey, CA 93942
(408) 646-6208

**Impostrip**
**Ultimate Technographics Inc.**
4980 Buchon St., Suite 403
Montreal, Canada, H4P 1S8
(514) 733-1188

**Microsoft Word/Microsoft Windows/**
**OS/2 Presentation Manager**
**Microsoft Corp.**
10611 NE 36th St.
Redmond, WA 98073
(206) 882-8080

**Per:Form**
**Delrina Technology**
10 Brentcliffe Rd., Suite 210
Toronto, Canada M4G3Y2
(416) 423-0456

**PS Tutor**
**Lincoln & Co.**
29 Domino Dr.
Concord, MA 01742
(617) 369-1441

**TOPS**
**Sitka Corp.**
950 Marina Village Parkway
Alameda, CA 94501
(415) 769-9669

## Scanners

**Array Scanner One**
**Array Technologies**
7730 Pardee Ln.
Oakland, CA 94621
(415) 633-3000

**Color Imaging System**
**BarneyScan Corp.**
1125 Atlantic Ave.
Alameda, CA 94501
(415) 521-3388

**LS-3500**
**Nikon Inc.**
623 Stewart Ave.
Garden City, NY 11530
(516) 222-0200

**MSF-300Z**
**Microtek Lab**
680 Knox St.
Torrance, CA 90502
(213) 321-2121

**ScanJet Plus**
**Hewlett-Packard**
19310 Pruneridge Ave.
Cupertino, CA 95014
(800) 752-0900

**3c**
**XRS Corp.**
1687 E. Del Amo Blvd.
Carson, CA 90746
(213) 608-3711

**TZ-3**
**Truvel Corp.**
8943 Fullbright Ave.
Chatsworth, CA 91311
(818) 407-0189

**UG80**
**Umax Technologies**
2352 Walsh Ave.
Santa Clara, CA 95051
(408) 982-0771

**Radius Two-Page Display**
**Radius, Inc.**
1710 Fortune Dr.
San Jose, CA 95131
(408) 434-1010

**VIP 640**
**Ventek Corp.**
31336 Via Colinas, Ste. 102
Westlake Village, CA 91362
(818) 991-3868

## Other Hardware

**Bernoulli Drive**
**Iomega Corp.**
1821 West 4000
South Roy, UT 84067
(801) 778-1000

**Color Keys/MatchPrint**
**3M Printing and Publishing**
**Systems**
3M Center Bldg. 223-2N-01
St. Paul, MN 55144
(612) 733-3497

**Cromalin**
**Du Pont Co.**
External Affairs Dept.
Wilmington, DE 19898
(302) 992-4217

**ColorScript Model 10**
**QMS Inc.**
1 Magnum Pass
Mobile, AL 36619
(205) 633-4300

**Pantone Matching Guide**
**Pantone, Inc.**
55 Knickerbocker Rd.
Moonachie, NJ 07074
(201) 935-5500

**Personal LaserWriter NT**
**Apple Computer**
20525 Mariani Ave.
Cupertino, CA 95014
(408) 996-1010

**PrecisionColor Calibrator**
**Radius, Inc.**
1710 Fortune Dr.
San Jose, CA 95131
(408) 434-1010

# INDEX

benefits of 21
files
  benefits of 75-76
  downloading 84-85
  outputting 76
  producing 80-84
  size of 76
  structure of 78
  transferring 77-78
font scaling 48
function of 18
output from Windows 35-36
printing options on Macintosh 31
printing to disk 81
RIPs in Linotronic imagesetters 14
support in Macintosh environment 28
PrePrint 154
Presentation Manager. *See* OS/2
Printing, offset. *See* Lithography
Printing plates 183
Printing press
  reproduction of photographs 96
  types of 181
Processor. *See* Film processor:

## Q

QMS ColorScript 144-145
QuarkXPress 153

## R

Radius 141
Raster Image Processor. *See* RIP
Registration
  marks 188
  process color separations 156-157
  repeatability problems 137
  spot color separations 129
Removable hard disks 171
Resolution
  adjusting color scanner 147
  halftone 97
  imagesetter options for outputting halftones 120
  laser printers 2
  Linotronic imagesetters 3
  Macintosh display 40
  of Linotronic models 17
  selecting via LCD panel 16
  specifying for service bureau 174
RGB
  color model 127
RIP
  function of 13-15
  interpretation of PostScript 73
Rotation
  of text 31

## S

Scaling
  scanned images 106-107
Scanners
  color 142-144, 146-148
  control software 105-112
  drum 131-132
  file formats 147-148
  image capture 101-107
  prescreened halftones 110
  preview scan 106
  slide 103
  types of 102-103
Screen angles. *See* Halftones: screen angles
SCSI
  interface for scanners 105
Selection
  tools in image software 116, 149
Separations. *See* Color: separations
Serifs 50-53
Service bureaus
  charges for film output 187
  color printer output 145
  description of 159
  reducing charges 197-202
  selecting 160-165
  sending PostScript files 76, 165-170
  submitting job ticket 174
  transporting files 170-174
Signatures. *See* Impositions
Slide scanners. *See* Scanners: types of
Slides
  producing on imagesetter 209-210
SmartArt 90
Standards
  IBM display 140
  in desktop color 135
Storage media
  for transporting files 171
Stripping
  impositions 191
  in lithography 182
Stuff-It 77, 168
Suitcase II 68-69
System 7
  outline fonts in 62

## T

Tabloid pages 169
TIFF
  as halftone format 147
  background on 110
TOPS 85
Transfer function
  adjusting in software 151
Trapping 156, C-4

Typefaces. *See* Fonts
Typesetting
  front-ends 11
Typesetting languages 6, 26

## U

Undercolor removal 157-158
Unix workstations 39

## V

Ventura Publisher
  extracting pages from PostScript file 205-207
  image control in 119
  producing PostScript output 81
  role of GEM 37
  sending files to service bureau 168

## W

Word processors
  viewing PostScript files 78